Medical Waste Incineration and Pollution Prevention

Medical Waste Incineration and Pollution Prevention

Edited by
Alex E. S. Green, Ph.D.
University of Florida, Gainesville

VNR VAN NOSTRAND REINHOLD
New York

Van Nostrand Reinhold
115 Fifth Avenue
New York, New York 10003

Chapman and Hall
2–6 Boundary Row
London, SE1 8HN, England

Thomas Nelson Australia
102 Dodds Street
South Melbourne 3205
Victoria, Australia

Nelson Canada
1120 Birchmount Road
Scarborough, Ontario M1K 5G4, Canada

16 15 14 13 12 11 10 9 8 7 6 5 4 3 2 1

Library of Congress Cataloging-in-Publication Data

Green, Alex Edward Samuel, 1919–
    Medical waste incineration and pollution prevention/Alex E.S. Green.
        p.   cm.
    Includes bibliographical references and index.
    ISBN 0-442-00819-8
    1. Infectious wastes.   2. Hazardous wastes—Incineration.
    I. Title.
    RA567.7.G74   1992
    363.72'88—dc20   92-958
                                                                    CIP

Contributors

Christopher D. Batich, Material Science Department, University of Florida, Gainesville, FL

Michael M. Bulley, Executive Director, Auckland Environmental Protection Agency Limited, Auckland, New Zealand

Daniel P. Y. Chang, Department of Civil Engineering, University of California, Davis, CA

Laura Constantine, Doucet & Mainka, P.C., Peekskill, NY

David Corbus, Private Consultant, Golden, CO

Harold Glasser, Department of Civil Engineering, University of California, Davis, CA

Alex E. S. Green, Clean Combustion Technology Laboratory, Department of Mechanical Engineering, University of Florida, Gainesville, FL

Floyd Hasselriis, Consulting Engineer, Forest Hills, NY

Donald C. Hickman, Captain, United States Air Force, Kirtland Air Force Base, NM

John C. Wagner, Clean Combustion Technology Laboratory, Department of Mechanical Engineering, University of Florida, Gainesville, FL

Kenneth B. Wagener, Department of Chemistry, University of Florida, Gainesville, FL

Contents

Preface

The annual cost of medical care in the United States is rapidly approaching a trillion dollars. Without doubt, much of the rise in costs is due to our health industry's concentration on high technology remediation and risk avoidance measures. From recent public discussions it is becoming increasingly evident that to contain the costs and at the same time extend the benefits of health care without national bankruptcy will necessitate much greater attention to preventative medicine.

The total cost of waste disposal by our health industry is well over a billion dollars. It is rising rapidly as we increasingly rely on high technology remediation measures. Here, too, in the opinion of the authors of this work, it would be prudent to give much greater attention to preventative approaches.

Incineration technology has largely been developed for disposing municipal solid waste (MSW) and hazardous waste (HW). As a result of the multibillion dollar funding for the Resource Conservation and Recovery Act (RCRA), most experts believe that pollution control is the key to minimizing toxic emissions from incinerators. This view is now beginning to take hold in medical waste (MW) incineration as well. However, the authors contributing to this book have concluded that precombustion measures can be most effective in reducing the toxic products of medical waste incineration.

We call this approach *pollution prevention*. *Toxic prevention* might be considered even more descriptive, since the Pollution Prevention Act of 1990 focuses on recycling, reduced packaging, and waste minimization in a number of industries, but not yet in the incinerator industry. However, the Clean Air Act Amendment(s) of 1990 define(s) 189 toxic substances to come under regulatory control in the next decade, in effect placing many toxic emissions from incineration in a similar regulatory category as the criteria pollutants defined in the Clean Air Act of 1970 (i.e., carbon monoxide, nitric oxides, sulfur dioxide, hydrocarbons, ozone, and particulates).

In organizing this book, I first attempted to follow the natural subdivisions of clean combustion technology as they have developed for many fuels (coal, oil, wood, and so forth). These may be defined as (1) precombustion measures, (2) combustion measures, (3) postcombustion measures, and (4) ash treatment and disposal. However, after drafts of several chapters were in my hands, it became clear that an emerging incineration–pollution prevention theme necessitated data spanning several subdivisions of clean combustion technology. Therefore, I have let certain redundancies and minor inconsistencies stand since they document how the different authors arrived at their complementary approaches and the minor differences that remain.

Whether the conventional professional wisdom in medical waste incineration can be altered to encompass vigorous precombustion measures is still an open question. Quite apart from the fact that experts disagree on the optimum thermal treatment of medical waste is the thorny problem of public perception. The latter too often means that policy debate on the subject is dominated by emotions rather than reason, particularly at the local level. In large part, emotion stems from public fears of the propagation of communicable diseases such as AIDS and hepatitis B via emissions or ash from medical incinerators. Yet the thorough destruction of infectious agents by fire and the breakdown of toxics are major advantages of incineration with respect to alternatives such as steam sterilization, microwave heating, or chemical treatment, all followed by landfilling.

By compiling the information herein, it is the authors' hope that our combined experience with precombustion controls will provide balance to the increasing emphasis on postcombustion (or pollution control) measures. A better understanding of what pollution prevention can accomplish will not only assuage public fears but also pave the way to the most cost-effective combination of clean combustion technologies for medical waste incineration.

I would like to thank the authors who have contributed chapters to the work. I would particularily like to thank Mike Bulley for sending me a copy of the recent Declaration by the International Union of Air Pollution Prevention Associations (IUAPPA), which follows. Last, but certainly not least, I would like to thank my wife, Freddie, whose retirement from her former position made it possible for me to complete this work in a timely fashion.

Alex E. S. Green

DECLARATION—The International Union of Air Pollution Prevention Associations (IUAPPA),

a non-governmental, non-political organization, consisting of professional and voluntary associations worldwide, whose national governments are assembling at the United Nations Conference on Environment and Development, respectfully submits to the United Nations and to all the governments of the world, for earnest consideration, the following declaration:

The UNION

—*Considering the full scope of the observations and recommendations for "sustainable development" contained in the 1987 World Commission on Environment and the Development report, titled "Our Common Future," to the United Nations General Assembly, including those focusing on the rate of human population growth in the world,*

—*Considering further the need for economic growth and agricultural production to feed this population, inevitably leading to more pollution,*

—*Cognizant of the great strides in environmental protection made over the last twenty years,*

—*Concerned that technology is reaching the limits of traditional methods of pollution treatment and control,*

—***Sensitive to the dilemma that the significant resources spent on pollution control are not available for improving productivity or implementing alternative pollution control measures,***

—***Recognizing that pollution control systems may result in pollutants being transferred from one medium to another,***

—*Concerned with the risks and potential social and environmental costs inherent in any release of pollutants into the environment,*

—*Aware that the public desires an environment where risks are minimized,*

—*Determined to promote the enhancement and maintenance of environmental quality, not only locally, but worldwide,*

SUBMITS, FOR THE PURPOSE OF THIS DECLARATION, THAT:

Pollution Prevention

—*Constitutes a cornerstone of sustainable development,*

—*Reduces the risk inherent in the management of some waste streams and residues that result from traditional control methods, including the risk of technology failure,*

—***Avoids the inadvertent transfer of pollutants across media that may occur with some "end-of-the-pipe" media-specific treatment and control approaches,***

—*Addresses certain environmental problems of extraordinary urgency, such as the perturbation of the earth's radiation balance with consequences for global climate changes,*

—*Applies to a broad array of activities that lead to pollution, including energy use, agriculture, transportation, as well as industrial activity,*

—*Protects natural resources for future generations, by avoiding excessive levels of wastes and residues, minimizing the depletion of resources, and maintaining the capacity of the environment to absorb pollutants,*

—**Provides a cost-effective method of environmental protection that can reduce raw material and energy losses, reduce the need for expensive "end-of-pipe" treatment technologies, encourage improvements in process efficiency and performance, and reduce long-term liability,**

—*Reduces the use of hazardous and toxic substances in manufacturing and other processes,*

—*Is feasible, practical, and available,*

DEFINES, FOR THE PURPOSE OF THIS DECLARATION, THE FOLLOWING CONCEPTS:

—**life cycle "cradle to grave" management includes raw materials extraction and use, energy conversion, impacts, transportation, worker safety, waste management, treatment and disposal, and potential liabilities, releases into the environment, as well as product use and ultimate disposal,**

—*support for better design of industrial processes includes research and development, technology transfer, economic incentives, and technical assistance,*

—*the sectors of the economy include production, energy (efficiency), product design, and renewable fuels,*

—*international ventures includes the promotion and dissemination by industrialized nations of low-polluting and low-waste technology in developing countries,*

—*public education includes providing information on consumer products and the consequences of business and industrial activity in their communities.*

CONCLUDES THAT:

—*Continued progress in environmental protection will require application of both innovative and traditional approaches for pollution control,*

—**Pollution prevention is the best possible solution for environmental protection on both environmental and economic grounds, being potentially the most effective method for reducing risks to human health and the environment and for containing costs,**

CALLS ON ALL GOVERNMENTS TO:

—*Orient their existing environmental programs to emphasize pollution prevention,*

—*Develop and use compatible analytical methods to assess the costs and environmental impacts of the entire life cycle management of products,*
—*Support the development and dissemination of better designs for industrial processes, inter-alia, to reduce the use of energy and scarce raw materials, and toxic pollutants, and the release of pollutants,*
—**Lead in the adoption of pollution prevention techniques through government procurement practices, the design and operation of government facilities, and the development of a mix of economic and regulatory incentives,**
—*Allow the maximum opportunity for flexibility and innovation in the design and pollution prevention approaches by industry and all other sectors of the economy,*
—*Support cooperative international ventures,*
—**Involve the public, as citizens and as consumers, in pollution prevention through education,**
—*Promote the use of pollution prevention impact statements, and*
—*Establish through an international forum, an appropriate demonstration of pollution prevention.*

Approved this 4th day of September, 1991, at Seoul, Korea, and signed on behalf of the Executive Committee of the International Union of Air Pollution Prevention Associations.

John Langston, Director General, IUAPPA
136 North Street, Brighton, BN1 1RG, UK

Medical Waste Incineration and Pollution Prevention

1

Toxic Products of Medical Waste Incineration

Alex E. S. Green and John C. Wagner

1.1 INTRODUCTION

Title III of the Clean Air Act Amendments of 1990 defines 189 toxic chemicals and 250 source categories that will come under regulatory constraints at various times in the 1990s (U.S. House of Representatives Report 1990). Table 1–1 lists the compounds that are deemed to present, through inhalation or other routes of exposure, a threat of adverse human health effects or adverse environmental effects. The health effects include, but are not limited to, substances that are known to be or may reasonably be anticipated to be carcinogenic, mutagenic, teratogenic, or neurotoxic; that cause reproductive dysfunction; or that are acutely or chronically toxic. Many of these chemicals have been detected in hospital incinerator tests by the California Air Resources Board (CARB) (Fry et al. 1990) and other agencies (see Table 1–2).

Hospitals employ toxic chemicals and hazardous materials for numerous diagnostic and treatment purposes (U.S. EPA 1990). The hazardous materials include (1) chemotherapy and antineoplastic chemicals, (2) formaldehyde, (3) photographic chemicals, (4) solvents, (5) mercury, (6) waste anesthetic gases, and (7) other toxic, corrosive, and miscellaneous chemicals. The volumes of these hazardous wastes are usually small compared to general hospital waste or red-bag waste (infectious material usually packaged in red-colored plastic bags). Accordingly, it is sensible to use special hazardous waste channels or waste minimization methods to handle these special wastes and *not* to place them in incinerator burn bags. Table 1–3 lists some waste minimization methods for general medical and surgical hospitals identified in this U.S. Environmental Protection Agency (EPA) study.

1

TABLE 1–1. Initial List of Hazardous Air Pollutants Established by the Clean Air Act Amendments of 1990

Acetaldehyde	DDE	Hexane
Acetamide	Diazomethane	Hydrazine
Acetonitrile	Dibenzofurans	Hydrochloric acid
Acetophenone	1,2-Dibromo-3-chloropropane	Hydrogen fluoride
2-Acetylaminofluorene	Dibutylphthalate	Hydroquinone
Acrolein	1,4-Dichlorobenzene(p)	Isophorone
Acrylamide	3,3-Dichlorobenzidene	Lindane (all isomers)
Acrylic acid	Dichloroethyl ether	Maleic anhydride
Acrylonitrile	1,3-Dichloropropene	Methanol
Allyl chloride	Dichlorvos	Methoxychlor
4-Aminobiphenyl	Diethanolamine	Methyl bromide
Aniline	N,N-Diethyl aniline	Methyl chloride
o-Anisidine	Diethyl sulfate	Methyl chloroform
Asbestos	3,3-Dimethoxybenzidine	Methyl ethyl ketone
Benzene	Dimethyl aminoazobenzene	Methyl hydrazine
Benzidine	3,3'-Dimethyl benzidine	Methyl iodide
Benzotrichloride	Dimethyl carbamoyl chloride	Methyl isobutyl ketone
Benzyl chloride	Dimethyl formamide	Methyl isocyanate
Biphenyl	1,1-Dimethyl hydrazine	Methyl methacrylate
Bis(2-ethylhexyl)phthalate	Dimethyl phthalate	Methyl tert butyl ether
Bis(chloromethyl)ether	Dimethyl sulfate	4,4-Methylene bis
Bromoform	4,6-Dinitro-o-cresol, and salts	Methylene chloride
1,3-Butadiene	2,4-Dinitrophenol	Methylene diphenyl
Calcium cyanamide	2,4-Dinitrotoluene	diisocyanate
Caprolactam	1,4-Dioxane	4,4'-Methylenedianiline
Captan	1.2-Diphenylhydrazine	Naphthalene
Carbaryl	Epichlorohydrin	Nitrobenzene
Carbon disulfide	1,2-Epoxybutane	4,-Nitrobiphenyl
Carbon tetrachloride	Ethyl acrylate	4-Nitrophenol
Carbonyl sulfide	Ethyl benzene	2-Nitropropane
Catechol	Ethyl carbamate	N-Nitroso-N-methylurea
Chloramben	Ethyl chloride	N-Nitrosodimethylamine
Chlordane	Ethylene dibromide	N-Nitrosomorpholine
Chlorine	Ethylene dichloride	Parathion
Chloroacetic acid	Ethylene glycol	Pentachloronitrobenzene
2-Chloroacetophenone	Ethylene imine	Pentachlorophenol
Chlorobenzene	Ethylene oxide	Phenol
Chlorobenzilate	Ethylene thiourea	p-Phenylenediamine
Chloroform	Ethylidene dichloride	Phosgene
Chloromethyl methyl ether	Formaldehyde	Phosphine
Chloroprene	Heptachlor	Phosphorus
Cresols/Cresylic acid	Hexachlorobenzene	Phthalic anhydride
o-Cresol	Hexachlorobutadiene	Polychlorinated biphenyls
m-Cresol	Hexachlorocyclopentadiene	1,3-Propane sultone
p-Cresol	Hexachloroethane	beta-Propiolactone
Cumene	Hexamethylene-1,6-diisocyanate	Propionaldehyde
2,4-D, salts and esters	Hexamethylphosphoramide	Propoxur (Baygon)

TABLE 1–1. (*Continued*)

Propylene dichloride	1,2,4-Trichlorobenzene	Arsenic Compounds
Propylene oxide	1,1,2-Trichloroethane	Beryllium Compounds
1,2-Propylenimine	Trichloroethylene	Cadmium Compounds
Quinoline	2,4,5-Trichlorophenol	Chromium Compounds
Quinone	2,4,6-Trichlorophenol	Cobalt Compounds
Styrene	Triethylamine	Coke Oven Emissions
Styrene oxide	Trifluralin	Cyanide Compounds
2,3,7,8-Tetrachlorodibenzo-	2,2,4-Trimethylpentane	Glycol ethers
p-dioxin	Vinyl acetate	Lead Compounds
1,1,2,2-Tetrachloroethane	Vinyl bromide	Manganese Compounds
Tetrachloroethylene	Vinyl chloride	Mercury Compounds
Titanium tetrachloride	Vinylidene chloride	Fine mineral fibers
Toluene	Xylenes	Nickel Compounds
2,4-Toluene diamine	o-Xylenes	Polycyclic Organic Matter
2,4-Toluene diisocyanate	m-Xylenes	Radionuclides (including
o-Toluidine	p-Xylenes	radon)
Toxaphene	Antimony Compounds	Selenium Compounds

Source: U.S. House of Representatives Report 1990.

TABLE 1–2. Compounds Detected in California Air Resources Board (CARB) Tests of Medical Waste Incinerators

Arsenic	1,2-Dibromoethane	Polycyclic aromatic
Ammonia	Dichloromethane	hydrocarbons
Benzene	Dichloroethane	Naphthalene
Bromodichloromethane	1,2-Dichloropropane	Sulfur dioxides
Cadmium	Dioxins/Furans	Tetrachloroethene
Carbon dioxide	Ethyl benzene	Tetrachloromethane
Carbon monoxide	Freon	Tetratrichloromethylene
Carbon tetrachloride	Hydrogen chloride	Toluene
Chromium	Hydrogen fluoride	Total hydrocarbons
Chromium^{+6}	Iron	Tribromomethane
Chlorodibromomethane	Lead	Trichloroethane
Chlorobenzenes	Manganese	1,1,1-Trichloroethane
Chloroform	Mercury	Trichloroethylene
Chlorophenols	Mesitylene	Trichloromethane
Copper	Methylisobutylketone	Trichlorotrifluorethane
Cumene	Nickel	Vinyl chloride
	Nitrogen oxides	m-, o-, p-Xylenes
	Particulate matter	Zinc

Source: Fry et al. 1990.

TABLE 1-3. Waste Minimization Methods for General Medical and Surgical Hospitals

Waste Category	Waste Minimization Method
Chemotherapy and antineoplastics	Reduce volumes used Optimize drug container sizes in purchasing Return outdated drugs to manufacturer Centralize chemotherapy compounding location Minimize waste from compounding hood cleaning Provide spill cleanup kits Segregate wastes
Formaldehyde	Minimize strength of formaldehyde solutions Minimize wastes from cleaning of dialysis machines and RO units Use reverse osmosis water treatment to reduce dialysis cleaning demands Capture waste formaldehyde Investigate reuse in pathology, autopsy laboratories
Photographic chemicals	Return off-spec developer to manufacturer Cover developer and fixer tanks to reduce evaporation, oxidation Recover silver efficiently Recycle waste film and paper Use squeegees to reduce bath losses Use countercurrent washing
Radionuclides	Use less hazardous isotopes whenever possible Segregate and properly label radioactive wastes, and store short-lived radioactive wastes in isolation on site until decay permits disposal in trash
Solvents	Substitute less hazardous cleaning agents, methods for solvents cleaners Reduce analyte volume requirements Use premixed kits for tests involving solvent fixation Use calibrated solvent dispensers for routine tests Segregate solvent wastes Recover/reuse solvents through distillation
Mercury	Substitute electronic sensing devices for mercury-containing devices Provide mercury spill cleanup kits and train personnel Recycle uncontaminated mercury wastes using proper safety controls
Waste anesthetic gases	Employ low-leakage work practices Purchase low-leakage equipment Maintain equipment properly to avoid leaks
Toxics, corrosives, and miscellaneous chemicals	Inspection and proper equipment maintenance for ethylene oxide sterilizers Substitute less toxic compounds, cleaning agents Reduce volumes used in experiments Return containers for reuse, use recyclable drums Neutralize acid waste with basic waste Use mechanical handling aids for drums to reduce spills Use automated systems for laundry chemicals Use physical instead of chemical cleaning methods

Source: U.S. EPA 1990.

Prior to 1990 the U.S. EPA's Waste Minimization Research Program (WMRP) focused on hazardous waste as defined in the Resource Conservation and Recovery Act (RCRA). The current Pollution Prevention Research Program (PPRP) expands U.S. EPA's WMRP to consider more than just RCRA hazardous waste (Freeman 1990). The goals of the PPRP program are as follows:

1. Stimulate the development and use of products that result in reduced pollution.
2. Stimulate the development and implementation of technologies and processes that result in reduced pollution.
3. Expand the reusability and recyclability of wastes and products and the demand for recycled materials.
4. Identify and promote the implementation of effective nontechnical approaches to pollution reduction.
5. Establish a program of research that will anticipate and address future environmental problems and pollution prevention opportunities.
6. Conduct a vigorous technology transfer and technical assistance program that facilitates pollution prevention strategies and technologies.

In a study of pollution prevention opportunities for the 1990s, Licis, Skovronek and Drabkin (1991) identify 17 industries where research could lead to attractive waste minimization opportunities but do not include our health industry. The major aim of this work is to further extend and develop our Pollution Prevention Program to encompass the lessons of clean combustion technology and to apply the combined thrust to medical waste disposal by incineration. As articulated in the phrase "purification by fire," incineration is the best method for dealing with the infectious component of the medical waste stream. Here we particularly wish to reduce toxic products of medical waste incineration, such as those listed in Table 1–2.

When the incineration method of medical waste disposal is used, not only is it important to minimize toxic materials (TMs) in the hospital waste stream, but it is also important to minimize toxic-producing materials (TPMs). Such materials might be benign in normal form but could produce corrosive or toxic substances in the high temperature chemistry of the incinerator's fire. Some familiar examples are the many products of incomplete combustion (PICs) and polynuclear aromatic hydrocarbons (PAHs) in the smoke of a smoldering or poorly burning wood fire. Such toxic hydrocarbon compounds have arisen in man-made or natural (lightning-induced) fires since time immemorial. The toxicity of such hydrocarbon compounds, however, is generally low compared to the toxicity of PICs

TABLE 1–4. Emission Rates from Medical Waste Incinerators

Compound	Emission Rate (lb/h × 10⁶)				
	Sutter General	Stanford	St. Agnes	St. Bernardines	Cedars Sinai
Dichlorofluoromethane	46	1.3	1.6	4.3	
Dichloromethane	390	160	550	34	
Trichlorofluoromethane	75	2.1	6.8	3.2	
Trichloromethane	ND	30	2.5	NM	
1,2-Dichloroethane	ND	170	1,600	2.2	
1,1,1-Trichloroethane	270	41	700	31	
Carbon tetrachloride	ND	2.9	28	1.9	52
Trichloroethene	61	11	550	0.76	12
1,2-Dibromomethane	ND	1.8	5.8	0.25	
Tetrachloroethene	280	10	170	11	56
Trichlorotrifluoroethane	130	10	27	4.7	22
Benzene	240	740	250,000	33	
Toluene	540	76	100	68	
Ethyl Benzene	150	17	40	3.4	
p-Xylene	120	140		4.2	
p + m-Xylene			3,000		
m-Xylene	280	120		37	
Cumene	ND	28	150	2.1	
o-Xylene	160	24	23	28	
Mestitylene	54	130	100	1.4	
Naphthalene	42	59	NM	21	
Methyl isobutyl ketone	ND	250–4	NM	5.1	
Dioxins (tetraocta)	7.9	0.037	3.9	0.078	1.3
Furans (tetraocta)	15	0.077	7.9	0.19	4.2
HCl (g/kg)	10	0.01	5.99	32.13	8.14
Feed rate (lb/h)	334/452	620	783	43/80	980

ND = not detected; NM = not measured.

Source: Adapted from California Air Resources Board and EPA documents.

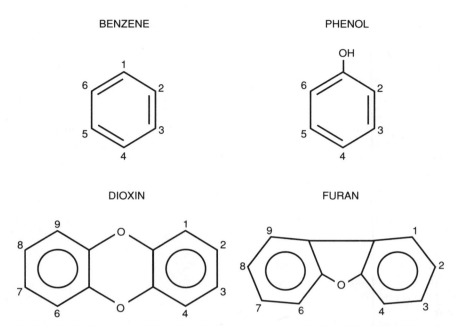

FIGURE 1–1. Structures of four classes of organic compounds of particular interest in toxic emission studies.

and PAHs of manufactured halogenated hydrocarbon materials. Accordingly, the minimization of toxic halogenated hydrocarbon compounds will be given special emphasis in this chapter and in later chapters of this work. The metallic compounds listed at the end of Table 1–1 constitute another important class of toxic compounds. Their origin and minimization will be discussed in later chapters.

1.2 TOXIC PRODUCTS OF INCINERATION

According to the Clean Air Act of 1970, "criteria pollutants" at concentrations of 1 to 1,000 parts per million (ppm) can result in health or environmental effects. Many of the toxics listed in Tables 1–1 and 1–2 can impact health at concentrations in the parts per billion or even parts per trillion range. Table 1–4 lists emission rates of hydrocarbons and chlorinated hydrocarbons measured in tests by CARB on medical waste incinerators (McCormack et al. 1988a,b; Jenkins et al. 1987b, 1988a,b). Figure 1–1 gives the generic molecular structures of benzene, phenol, dioxin, and furans—toxic chemicals that have been of particular interest in toxicity studies related to incineration. In normal hydrocarbon forms, these aromatic compounds have hydrogen atoms in the numbered positions. When these positions are occupied by chlorine atoms, the various *congeners* are designated by the occupied positions and the labels for the molecular class. For example, for twice-chlorinated benzene, 1,2-dichlorobenzene, 1,3-dichlorobenzene, 1,4-dichlorobenzene, etc. are isomers having the same molecular weight. However, their chemical properties and toxicities can vary greatly.

The words *dioxins* and *furans* are frequently used to denote polychlorinated dibenzo-*p*-dioxins (PCDDs) and polychlorinated dibenzofurans (PCDFs) (see Figure 1–1). Specific PCDDs and PCDFs are identified by the number and specific placement of chlorines in the structure. In unchlorinated dibenzo-*p*-dioxin, all labeled positions are occupied by hydrogen atoms. A particular congener of this compound is 2,3,7,8-tetrachloro-dibenzo-*p*-dioxin, or 2,3,7,8-TCDD (or -TeCDD). This congener is the most toxic PCDD and is generally used as the standard of toxicity. Table 1–5 lists the relative toxicities that have been assigned to various dioxin and furan congeners (Manscher 1989a; Jenkins et al. 1988b). Most of these relative assignments have been based upon mice experiments, and there is still considerable controversy as to the transferability of these results to humans (Hanson 1991).

A number of recent state-of-the-art reviews have been written on medical waste incineration (Cross et al. 1990; Theodore 1990; Barton et al. 1989; U.S. EPA 1988; Doucet 1991; Lauber and Drum 1990). The conventional wisdom that emerges in most of these works is the recommendation to incorporate postcombustion systems, i.e., pollution control devices, as the principal means of minimizing toxic emissions. As noted previously, the major aim of this book is to balance this trend by an exposition of what can be accomplished with pollution prevention or precombustion measures, the first law of clean combustion technology.

Whether the conventional professional wisdom of the day can be altered to encompass emphasis on pollution prevention is still an open question. Quite apart from the problem of professional disagreement on the best approach to the thermal treatment of medical waste, major public perception problems have reached the point where emotions and fear of infectious diseases rather than reason or quantitative data usually dominate public policy discussions. Yet the destruction of infectious agents by fire is the major advantage of the incineration method of medical waste disposal as compared to alternative treatments such as steam sterilization, microwave heating, chemical treatment (all followed by landfilling).

In the next section we will describe the considerations that originally led us to pursue precombustion measures together with advanced combustion techniques as essential components of institutional waste incineration.

TABLE 1-5. Toxic Equivalent Factors[a] According to Various Systems

		SCAND.[b]	EPA[c]	Eadon[d]	CARB[e]
Mono, di, tri	CDD	0.0	0.0	0.0	0.0
2,3,7,8	TeCDD	1.0	1.0	1.0	1.0
Other	TeCDD	0.0	1.0	0.0	0.0
2,3,7,8-subst.	PeCDD	0.5	0.5	1.0	1.0
Other	PeCDD	0.0	0.005	0.0	0.0
2,3,7,8-subst.	HeCDD	0.1	0.04	0.03	0.03
Other	HeCDD	0.0	0.004	0.0	0.0
2,3,7,8-subst.	HpCDD	0.01	0.001	0.0	0.03
Other	HpCDD	0.0	0.00001	0.0	0.0
	OCDD	0.001	0.0	0.0	0.0
Mono, di, tri	CDF	0.0	0.0	0.0	0.0
2,3,7,8	TeCDF	0.1	0.1	0.33	1.0
Other	TeCDF	0.0	0.001	0.0	0.0
1,2,3,7,8	PeCDF	0.01	0.1	0.33	1.0
2,3,4,7,8	PeCDF	0.5	0.1	0.33	1.0
Other	PeCDF	0.0	0.001	0.0	0.0
2,3,7,8-subst.	HeCDF	0.1	0.01	0.011	0.03
Other	HeCDF	0.0	0.0001	0.0	0.0
2,3,7,8-subst.	HpCDF	0.01	0.001	0.0	0.03
Other	HpCDF	0.0	0.00001	0.0	0.0
	OCDF	0.001	0.0	0.0	0.0

[a] Based on 2,3,7,8-TeCDD equal to 1.
[b] Nordic equivalent.
[c] U.S. EPA 86.
[d] Eadon-83, Denmark.
[e] CARB 1989.

Source: Adapted from McCormack 1988a.

1.3 CLEAN COAL TECHNOLOGY

In 1979 the United States experienced three energy shocks: (1) the Three-Mile Island accident, (2) the Iran revolution, and (3) a twofold increase in OPEC oil prices. That same year the Interdisciplinary Center for Aeronomy and (other) Atmospheric Sciences (ICAAS), established in 1970 at the University of Florida to address air pollution problems, was awarded funds to study the impact of increased coal use in Florida. At that time Florida lagged behind only California and New York in the use of oil in electricity generation and had a large number of relatively new oil boilers. More than 30 faculty members and students from many disciplines participated in this study. The conclusion in late 1980 (Green 1980) of this interdisciplinary effort was that we must decrease oil imports by conservation and by greater reliance on our domestic coal supplies burned in environmentally acceptable fashions.

Late in 1980 it became apparent that our reserves of domestic natural gas were not as low as prior official government policy had assumed. The Clean Combustion Technology Laboratory (CCTL), an offshoot of ICAAS, then initiated a research and development program on using gas to assist in the clean and efficient burning of coal in boilers designed for oil (Green 1981). During the course of this work we concluded that gas could also assist the burning of wood, municipal waste, sludge, and other potential domestic energy sources (Green 1986; Green et al. 1985, 1987, 1988). The concept of cofiring of gas and coal did not get a friendly reception. By advocating using some gas to make coal burn more cleanly and efficiently in oil boilers, we entered the crossfires of the fuel wars and at the same time violated conventional wisdom and governmental policy, which was to use coal to make oil and gas (our Synthetics Fuel Program). Nevertheless, after a number of scientific and engineering studies on natural gas–coal cocombustion (Green and Green 1981; Green et al. 1986), we reached the conclusion that synergisms ($1 + 2 = 4$) can be achieved in gas-solids cocombustion (Green and Pamidimukkala 1984).

When appointed a member of the National Coal Council, Green participated in an early study of clean coal technology (Blackmore and Leibson 1986). It became apparent that the general framework that has emerged for clean coal technologies is also applicable to other fuels, including biomass, solid waste, sludge, and other potential domestic sources of energy that might replace imported oil. The fact that an intense effort and a high level of funding (several billion dollars to date) have been applied to clean coal technology, whereas miniscule funds by comparison have been applied to biomass and solid waste, makes it important not to overlook the lessons learned from our national Clean Coal Technology Program.

The clean coal technology framework is usually divided into precombustion, combustion, and postcombustion measures. Included in the precombustion measures are coal cleaning by physical, chemical, or biological techniques to reduce sulfur or ash, which cause harmful emissions or inefficient burning. Another precombustion measure is to replace high sulfur coal by low sulfur coal. Among the combustion measures are included arrangements that minimize the generation of NO_x and help in the capture of SO_2. The postcombustion measures (see Chapter 6) include wet scrubbing, dry scrubbing, the use of advanced baghouses, cyclones, or precipitators to capture fly ash, sulfur oxides, and other harmful products of combustion. Various combinations of these measures can also be used. Table 1–6 is a list of clean coal technology approaches identified in 1985 for the purposes of retrofitting and converting industrial oil boilers while meeting new source performance standards (NSPS).

TABLE 1–6. Clean Coal Technology, Status of Various Options and Commercial Ready Dates

Precombustion		
Coal selection	C	1986
Physical cleaning	C	1986
Chemical cleaning	P	1995
Biological cleaning	P	1995
Combustion/Conversion		
Micronized coal	P	1990
Coal-Water slurry	P	1990
Coal-Water-Gas	P	1987
Two-stage slagging combustion	P	1992
Two-stage dry combustion	P	1988
Sorbent injection	P	1992
Low NO_x combustion	D	1988
Noncatalytic reduction of NO_x	D	1990
Atmospheric FBC	D	1990
Pressurized FBC	P	1992
Gasification combined cycle	D	1988
Gasification fuel cell	L	1998
Postcombustion		
Conventional wet FGD	C	1986
Spray dry FGD	C	1986
Advanced wet FGD	D	1990
Dry injection FGD	P	1990
Combined SO_x/NO_x removal	P	1995
Catalytic reduction of NO_x	D	1990
Advanced baghouse	P	1988
Advanced cyclones	L	1990
Advanced precipitators	L	1990

C = commercial; D = demonstrated; P = pilot plant;
L = laboratory.

Source: Adapted from Blackmore and Leibson 1986.

In retrospect, it should be noted that the 1985 clean coal technology framework focused mainly on reduction of criteria pollutants. It did not consider global anthropogenic emissions, which might impact the stratospheric ozone layer or enhance the earth's greenhouse effect. Nor did it consider toxic emissions. These issues became very topical after 1985. Figure 1–2 illustrates anthropogenic emissions that are currently of concern in energy, environmental, and economic discussions. The clean coal technology framework also did not explicitly address the solid residues of coal combustion (bottom ash, fly ash), however, these were and are implicitly considered. Thus, in addition to the precombustion, combustion, and postcombustion categories, which primarily address gaseous products, it

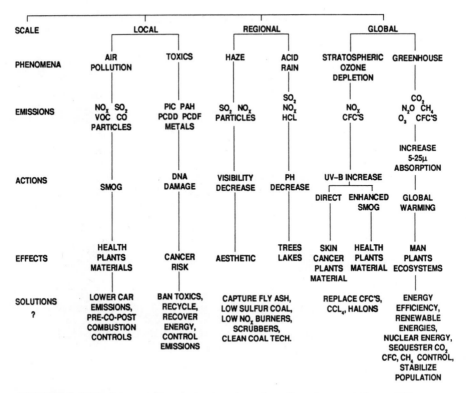

FIGURE 1–2. Anthropogenic emissions to atmosphere from transportation, utilities, industrial, commercial, residential, landfill, incineration, and agricultural sources. (Adapted from Green et al. 1990b.)

is appropriate to add residue disposal. The last category should consider all solid and possibly liquid waste products generated in the use of coal from mining operations to the production of electricity.

The opportunity to demonstrate potential synergisms in an industrial-size water tube oil boiler came in 1985 with industrial and university funding and with the assignment to the program of a 535 HP watertube boiler at the Sunland Training Center (now called Tacachale), which is approximately 4 miles from the University of Florida campus. Our modifications of the Tacachale boiler to serve as a quantitative calorimeter and our results on coal-water-gas cofiring have been reported at a number of technical conferences (Green et al. 1986). In addition to our experimental work, we devoted considerable attention to modeling the pyrolysis and combustion of coal-water slurries cofired with natural gas (Green and Pamidimukkala 1982). Unfortunately, funding for our work ceased after

TABLE 1–7. Clean Waste Combustion Technology

Precombustion
Label then ban use of hazardous compounds in everyday waste stream
Manufacture and use products to minimize waste
Sort recoverables, recyclables, compostibles, and feedstockibles
Use processors, flailers, sorters, driers to improve burning qualities and reduce hazardous
   waste generation

Combustion
Use gas-assist to compensate for fuel quality and hazardousness
Use multistaging with high temperature stages
Recover combustion energy
Use combination techniques and air and gas staging

Postcombustion
Hot gas cleanup
Advanced baghouses
Advanced precipitators
Use caustic scrubber to reduce acid emissions
Use residuals where possible
Properly landfill remainder

Source: Green 1986.

oil prices declined in 1986 from $30 per barrel to $10 per barrel and interest
in alternative fuels declined more precipitously.

Just as this study of cocombustion of coal-water slurry with natural gas
came to an end, we acquired, by donation, a 500 lb/h Environmental
Control Products incinerator, which formerly served the Veterans Hospital in Gainesville, Florida. In consideration of the lessons learned while
pursuing our coal-water-gas effort and while assisting in the clean coal
technology effort (see Table 1–6), we endeavored to apply these lessons
to clean waste combustion technology. Table 1–7 was our first translation
to waste of Table 1–6 for coal. In the next section, we will summarize the
experimental program on institutional waste disposal by the University
of Florida–Tacachale–Clean Combustion Technology Laboratory (UF-T-
CCTL).

1.4 THE UF-T-CCTL
EXPERIMENTAL PROGRAM

Sociotechnical Studies

In our pursuit of institutional solid waste disposal by combustion, we
decided to test the hypothesis that precombustion cleanup, by restriction
of the institutional use of disposable materials and by source separation

of TPMs, could substantially reduce the output of toxic combustion products. We also decided to test the hypothesis that an existing modular combustor with minor improvements in the form of retrofits (e.g., stokers, extra blowers, feedback controls) could in conjunction with precombustion measures reduce toxic products to levels now only achievable with highly expensive postcombustion controls. The fuels tested included the disposable nonhazardous waste (NHW) from Tacachale, in which paper and plastics are represented in even greater proportions than in medical waste. As is usual with medical incinerators, we cofired this NHW with natural gas (NG) when necessary. In connection with a program on the use of renewable energy, we conducted cofirings with cellulosic biomass (CB).

Tacachale, an institution for developmentally disabled persons, is organized into facilities consisting of groups of homes, each having 14 to 18 beds. Supervisors were instructed to divide waste into (1) toxic or hazardous components, (2) food waste, (3) recyclables, and (4) nonhazardous components. The list of toxic materials was condensed from a more comprehensive list (Tufts University 1987), which was subdivided into household, automotive, hobby product items, and hazardous materials such as batteries, aerosol cans, and bottled gas. In the initial experiments the NHW was bagged in 30-gallon transparent polyethylene bags by the staff and placed in CCTL dumpsters for weekly collection by CCTL personnel.

A motivational system was established to discourage placing TPMs, wet food waste, bottles, cans, and other recyclables in the burn bags and to encourage discarding only good, combustible material in NHW burn bags. The NHW bags were evaluated, upon weighing and transfer to the conveyor, as to packaging qualities, absence of cans and bottles, or apparent toxic or hazardous components. Scores for each home were posted after every trial burn. A dumpster painted silver was placed at the home with the highest score (The Silver Dumpster Award). With the passage of time, the quality of combustible material in our burn bags improved substantially. A source characterization in March 1990 based upon six representative burn bags identified 42 percent paper and cardboard, 39 percent plastics, 14 percent food, 2 percent cloth, 2 percent metal, and 1 percent glass. The plastics, mostly disposable items related to food service, consisted mainly of paper coatings and high impact polystyrene and very little polyvinyl chloride (PVC). Thus the Tacachale NHW simulated a potential future medical waste stream in which TPMs were avoided by the purchasing agent and by source separation at each ward.

In May 1990 the waste collection protocol for trial burns was changed from using the weekly waste of eight selected homes to using one day's waste from all 48 homes. The waste quality suffered with this change, and a new motivational system has been under development to correct this.

Incinerator

The two-chamber, ram-fed 500 lb/h Environmental Control Product incinerator, which was originally installed at the Veterans Hospital in Gainesville in 1973, was donated to the CCTL and delivered to the Tacachale steam plant in spring 1987. The incinerator was rebuilt, repaired, installed, and made operational by December 1987. The layout of the major components of the incinerator is shown in Figure 1–3. The trailer housing the data acquisition and continuous monitoring systems as well as the Florida Department of Environmental Regulations stack sampling truck are indicated. The sampling platform was designed and constructed to facilitate stack emission measurements at various points along the refractory flue venting to the chimney. A conveyor belt is used to transport NHW into the hopper for the incinerator ram feeding system. To minimize processing costs of CB, a versatile biomass feeding system was developed to handle CBs having a variety of textural qualities (Green et al. 1990a). The CB injection system consists of an 8″ × 10″ opening in the top of the primary combustion chamber, with a windbox and external fan providing overpressure locally to educe the chips inward.

Data Acquisition System and Instrumentation

The major components of the data acquisition system located in the instrument trailer are a Zenith Z-158 personal computer and a Doric Dataport 236 data logger. Seven type-K thermocouples are connected to the data logger through a front end module (FEM) installed in one of the incinerator's control panels. Various thermocouples are installed in the stack and the incinerator's chambers, and one is installed in a control panel to measure ambient temperature. The natural gas flow rate is measured by the gas company meter. Input air flow rates are measured by orifice meters. The computer automatically and continuously monitors the thermocouple, O_2, CO_2, and CO readings, and incinerator control settings, recording the data once every 11 seconds to a floppy disk. Three continuous emission monitors are installed inside the instrumentation trailer: a Beckman 864 CO_2 monitor, a Beckman 866 CO monitor with a zero/span accessory, and a Servomex 777 Combustion Analyzer, which monitors O_2. A second sampling line and a second monitor is available to monitor CO in the primary combustion chamber.

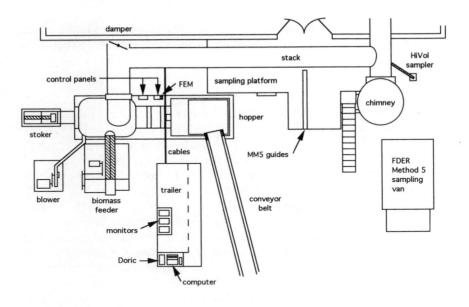

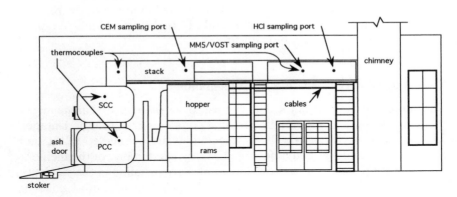

FIGURE 1–3. Clean combustion technology laboratory research incinerator. (From Green et al. 1990a.)

Stack Sampling

The CCTL has developed the capability to execute stack sampling for particulate concentrations by Method 5 (U.S. EPA 1979), for semivolatile organics by Modified Method 5 (U.S. EPA 1986a), for volatile organic compounds by VOST (Hansen 1984; U.S. EPA 1986b), and for HCl (U.S. EPA 1989). The Modified Method 5 samplings, the Soxhlet extraction, and the Kuderna Danish concentration procedures were carried out in-house. The gas chromatograph/mass spectrometer (GC/MS) analyses of the concentrates were carried out in the Analytical Instrument Laboratory of the Chemistry Department. We detected hydrocarbons, fatty acids, phthalates, alcohols, esters, aromatic hydrocarbons, and carbolic acids at levels of approximately 10 ppb. However, no organochlorine compounds were detected. In later attempts we extended our run time and hence sample volume. However, limitations on the total waste that our sociotechnical procedures provided kept us below the necessary collection volumes.

Because of this problem we focused on VOST measurements, which require only 20 minutes for an adequate sample. Here the preparation of the Tenax traps and Tenax-charcoal traps were carried out in house. The analysis was carried out with a GC/MS system in the Environmental Science and Engineering Department, which we adapted for VOST analyses. Our HCl measurements require 30- to 60-minute sampling times, which also permitted comparative experiments. Analyses of our samples were carried out with analytical equipment in the Soils Department.

Results of VOST and HCl Measurements

The CCTL has obtained reasonable measurements of HCl and volatile organic compounds (VOCs). These measurements include a number of chlorinated organic compounds that are considered to be precursors of chlorinated dioxins or furans. The volatile organic sampling employs water-chilled condensers to cool the stack gas and two sorbent cartridges (front and backup) to trap organic compounds. The front trap is the primary collector and was filled with about 1.6 g of Tenax-GC or Tenax-TA polymer resin. The backup trap contains about 1.0 g each of Tenax and charcoal. The sampling procedure requires 20 liters of stack gas to be pulled through the train in 20 minutes. Analysis was performed by thermally desorbing the cartridges into a purge and trap device and then injecting into a GC/MS. The success of these measurements is directly attributable to this sampling and analysis methodology, which effectively concentrates VOCs from a 20-liter stack gas volume into a single GC/MS analysis.

The GC/MS analysis typically targets U.S. EPA Priority Pollutants. Internal standard recoveries in all of our reported data were adequate. The second and third columns of Table 1–8 show the averaged emission rates obtained in eight trial burns of representative NHW and five trial burns using NHW spiked with PVCs. It appears that chloromethanes and chlorobenzenes increased significantly with PVC spiking. However, the chloroethanes and chloroethenes do not show as great a trend. Aromatic hydrocarbons tend to show increase emission rates with PVC spiking.

TABLE 1–8. Comparison of Results from CCTL Volatile Organic and Hydrogen Chloride Sampling with CARB Sampling

	CCTL Average	CCTL PVC-Spiked Average	California Hospital Average	CCTL/Cal Effective Scrubbing Efficiency
Feed Rate (lb/h)	356.8	351.0	558.3	
HCl (g/kg)	2	12	17	88%
Compound		Emission Rates (μg/kg)		
Benzene	143.0	419.2	80,501	100%
Toluene	128.1	93.8	891	86%
Ethylbenzene	21.4	38.2	146	85%
m-, p-Xylene	19.2	17.0	1,614	99%
o-Xylene	4.6	6.3	315	99%
1,2-Dichloroethane	0.0	0.0	593	100%
1,1,1-Trichloroethane	14.1	14.6	634	98%
Carbon tetrachloride	18.6	30.2	29	35%
Trichloroethene	1.9	5.0	184	99%
Tetrachloroethene	2.0	2.8	272	99%
Chloroform	16.1	65.5	17	5%
Chlorobenzene	62.0	169.5		
1,2-Dichlorobenzene	21.4	110.4		
1,3-Dichlorobenzene	17.7	76.5		
1,4-Dichlorobenzene	2.1	7.3		
1,1-Dichloroethane	6.2	0.0		
1,1-Dichloroethene	0.9	0.3		
1,1,2-Trichloroethane	0.0	3.8		

Source: Green et al. 1991c.

The fourth column of Table 1–8 gives the average volatile organics and hydrogen chloride emission rates given in reports of the CARB (see Table 1–4). The first four incinerators were without controls. In the last case measurements before the control system were used. In practically all cases the concentrations in our stack gases were substantially lower than the average concentrations of the same VOCs measured by CARB when normalized to the same feed rate. The fifth column of Table 1–8 gives the effective scrubbing efficiency achieved by CCTL with respect to the average of the five California incinerators.

It is notable that chloroethanes and chloroethenes measured by the CCTL are about two orders of magnitude lower than those measured by CARB. The aromatic hydrocarbons show similar trends. However, chloromethanes, although somewhat lower, appear to be similar in magnitude. Overall, the CCTL measured an order of magnitude less organically bound chlorine than were measured by CARB.

We attribute the lower toxic emissions indicated in the first CCTL columns of Table 1–8 to the following:

1. Our retrofit measures: (a) installation of a stoker, (b) incorporation of a strong blower for extra underfire air, (c) installation of an extra tangentially directed overfire air blower (originally for use as an educing agent for our biomass feeder), and (d) our operational protocol of running at higher temperatures rather than conventional starved air incinerator temperatures.
2. Our protocols of (a) separating toxics and recyclables and bagging only nonhazardous waste, (b) avoidance of halogenated plastics by the Tacachale purchasing agent, and (c) additional screening by incinerator personnel.

The fact that the CCTL trial burns show lower volatile chlorinated hydrocarbons (VCHCs) with respect to the average CARB trial burns and also lower VCHCs with respect to PVC-spiked CCTL burns is very convincing physical and chemical evidence that pollution prevention can work. These results raise the issue as to the origin of the conventional wisdom, which may be expressed in these two statements:

1. There is no evidence that the amount of PVC in the waste affects the levels of CHCs in the stack emissions.
2. There is no statistically significant relationship between the levels of CHCs and HCl emissions.

These statements were essentially given in the conclusion of the widely quoted Pittsfield-Vicon study (Neulicht 1987) but with the specific abbreviations PCDDs/PCDFs in place of our CHCs. This apparent physical dilemma together with our own inability to experimentally measure PCDDs/PCDFs motivated us to initiate efforts to also use data in the literature to determine whether relationships exist between CHCs and HCl emissions.

1.5 REANALYSIS OF CHC DATA

The CARB emission tests constitute one of the best sets of CHC emission data now available on medical incinerators operating near their normal operating conditions. These reports have been of particular interest since CARB has instituted regulations on dioxin emissions, whose effect was projected to lead to the shutdown of 135 of 146 medical incinerators in California (Fry et al. 1990). CARB uses dioxin as a generic name for a family of 75 PCDDs weighted according to toxicity in a standard way.

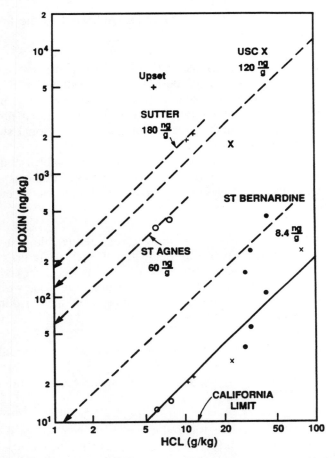

FIGURE 1–4. California Air Resources Board dioxin data vs. HCl data. (From Green et al. 1991b.)

Green et al. (1991b), on the basis of dioxin-HCl pairs assigned in CARB tests, examined the possibility that the ratio

$$M = \frac{Y}{X} \tag{1–1}$$

is a useful figure of merit (FOM) that characterizes the incinerator and its operating conditions. Here Y is the dioxin emission factor (in ng emission/kg waste input) and X is the HCl emission factor (in g/kg). Figure 1–4 shows the CARB data and the FOM lines (dashed) assigned by Green et al. The very large differences in FOM from 180 ng/g (Sutter) to 8.4 ng/g (St. Bernardine) are noteworthy, reflecting the construction of the incinerator and its mode of operation. The lower solid line and the translated points associated with it correspond to a hypothetical incinerator operating with an FOM of 2 ng/g. Equation 1–1 affords a plausible description of the limited dioxin data from medical incinerators during routine operations. The fact that FOM = M can vary greatly between incinerators or operating conditions is an important lesson.

To our knowledge the best set of controlled experiments reported in the literature for a single modular incinerator was carried out at the Vicon-Pittsfield incinerator with the support of the New York State Energy Research and Development Authority (Neulicht 1987). The Midwest Research Institute (MRI) has attempted to organize the Vicon-Pittsfield incinerator emissions by linear regression analysis. These results indicated strong correlations between some measured variables and weak and nonsignificant correlations between others. Green et al. (1991b) have reexamined the Vicon data seeking nonlinear relationships between the various measurables. They first examined a one-dimensional (1D) model that related chlorinated hydrocarbon emissions to HCl emissions as a power law, which may be expressed in the form

$$Y_i = a_i^2 \left(\frac{X}{X_u}\right)^{n_i}$$

(1–2)

Here Y_1 = PCDF = F, Y_2 = PCDD = D, Y_3 = ClBz = B, Y_4 = ClPh = P, X is the HCl concentration, and X_u is a convenient unit. The parameters a_i and n_i were obtained by least square minimization. These functions, when examined in relationship to the data, are not very impressive. However, most of the relationships between Y_is derived from Eq. 1–2 by eliminating HCl are quite impressive. This observation led to the examination of various two-dimensional (2D) $Y_i(HCl,T)$ and three-dimensional (3D) $Y_i(HCl,T,CO)$ functions, where T is the temperature and CO the carbon monoxide concentration at the tertiary duct (after the secondary combustion chamber). Among approximately 50 analytical functions examined, the best found to date is Model 41:

$$Y_i = a_i^2 X \exp n_i X + b_i^2 X [\exp m_i X] F(t;p,\mu) W^{q_i}$$

(1–3)

where W = CO/10, t = $(T + 460)/1,000$, or a scaled absolute temperature, and $F(t;p,\mu)$ is an Arrhenius reaction rate–type function:

$$F(t;p,\mu) = At^\mu \exp - (p/t) = A_o(t/2)^\mu \exp [(p/2) - (p/t)]$$

(1–4)

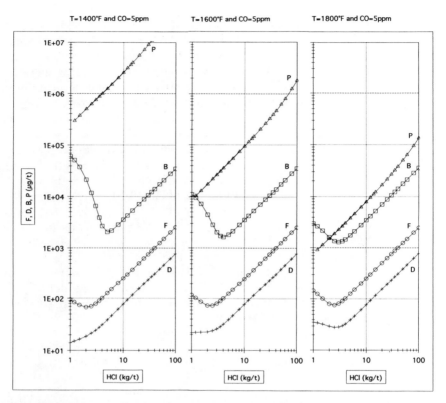

FIGURE 1–5. CHC emission factors vs. HCl emission factor according to Eq. 1–3 for T = 1,400°F, 1,600°F, and 1,800°F with CO = 5 ppm at tertiary duct.

In the second form of the Arrhenius function, t is normalized by a median value $t = 2$, which makes the amplitude factor A_o relatively insensitive to variations of the Arrhenius parameters p and μ. Figure 1–5 illustrates the nonlinear 3D fit for CO = 5 ppm for T = 1,400°F, T = 1,600°F, and T = 1,800°F.

The general noisiness of stack emission data when the measured species concentrations are in the parts per billion or parts per trillion is a serious problem in phenomenological modeling. For example, pairs of Vicon measurements that were supposed to represent duplicate conditions sometimes differ by an order of magnitude or greater. In attempts to overcome this difficulty, the CCTL group has examined a variety of data smoothing and data rejection prescriptions. The problem is illustrated in Table 1–9, which gives the original $Y_i(X, T, W)$ data at the tertiary duct extracted from the Vicon report. Lowercase letters denote pairs of data sets that were supposed to represent duplicate conditions. Vertical lines denote pairs that differ by large factors.

The possibility exists that variables other than X, T, and W may influence the outputs Y_i. Table 1–9C lists such potential variables. The most noteworthy data here are the very large values of PCDF and PCDD in the input waste, indicated by asterisks. The fact that these are for the PVC-free cases may be crucial to the failure of MRI to see correlations between PVC or HCl and PCDF or PCDD. Another problem that makes it difficult to pin down the optimum functions of $Y_i(X, T, Z)$ is the relatively good correlation between temperature and carbon monoxide in the Vicon data (Green et al. 1992).

Equation 1–3 lends itself to a reasonable physical interpretation consistent with many incinerator measurements in California, Canada (Clement et al. 1989), Denmark (Manscher et al. 1989), Britain (Brown et al. 1989), and other areas. In effect, the second term becomes small when the incinerator is operated at near-optimum conditions. The first HCl-dependent term may then represent the intrinsic limiting performance of the unit.

It is now generally recognized that the cooling process in the boiler creates additional dioxins and furans. However, under good combustion conditions one would expect all complex organic compounds exiting the secondary combustion chamber to be destroyed, leaving only simple inorganic gases. Under these circumstances, the formation of chlorinated aromatic organics during cooling in the boiler should be minimal. With these reasonable assumptions our analytic modeling of chlorinated organics at the duct before entering the boiler should be directly relevant to an overall combustion control system for toxic minimization by the complete combustion system.

To make sure that the empirical equations chosen for investigation did not reflect personal prejudice, the CCTL group also used linear multivariate statistical analysis on both the tertiary duct and boiler exit data. The objective was to test the dependence of Ys upon temperature, carbon monoxide, and HCl and the variables listed in Table 1–9B. They examined many relationships of the form

$$Y = a_0 + a_1 t + a_2 t^2 + a_3 Z + a_4 X + a_5 h + a_6 Cl + a_7 O_2$$
$$+ a_8 f + a_9 d + a_{10} w \tag{1–5}$$

where h, Cl, O_2, f, d, and w denote H_2O, Cl, excess O_2, furans, dioxins, and feed rate in the inputs. While in several cases the analyses showed

TABLE 1-9. Pittsfield-Vicon Data at Tertiary Duct

(A) Independent and Dependent Variables

	Run No.	Test Condition	HCl (kg/t)	T (°F)	CO (ppm)	PCDF (μg/t)	PCDD (μg/t)	ClBz (mg/t)	ClPh (mg/t)
a	11	MSW	3.40	1,327	152	380	160	2.5|	11.0
	22	MSW	3.21	1,283	251	630	220	7.9|	6.2
b	28	MSW	3.01	1,458	43	68	15	1.2	4.4
c	10	MSW	2.85	1,539	8	49	18	1.5	32.0|
	16	MSW	3.67	1,636	11	71	11	1.6	5.2|
d	15	MSW wet with rain	1.88	1,643	8|	87|	27|	7.9	5.2
	21	MSW + H$_2$O	3.99	1,508	25|	200|	110|	3.6	3.9
f	23	MSW – low O$_2$	13.79|	1,933	3|	340	100	4.6|	6.4|
	26		2.67|	1,834	12|	130	44	1.1	3.0|
h	12	PVC-free	0.52	1,796	0.5	51|	12	5.6|	14.0
	17		0.46	1,861	2	220|	11	2.2|	6.9
i	13	PVC-free + PVC	2.57	1,739	4	68	30	2.8	9.8|
	19		2.10	1,840	7	52	27	1.2	4.4|
j	24	PVC-free + H$_2$O	0.90	1,810	11	450|	110|	3.6|	4.5
	29		1.15	1,739	6	74|	50	1.1|	2.3

(B) Best Fit Parameters ($n = 0$, $\mu = 1$)

	$Y_1 = F$ = PCDF	$Y_2 = D$ = PCDD	$Y_3 = B$ = ClBz	$Y_4 = P$ = ClPh
alpha	5.012 ± 0.370	2.783 ± 0.274	0.594 ± 0.084	0.453 ± 0.586
beta	43.290 ± 23.962	11.175 ± 6.070	9.364 ± 4.136	2.528 ± 0.318
m	–2.079 ± 0.439	–1.368 ± 0.311	–1.756 ± 0.473	0.007 ± 0.102
p	3.654 ± 13.006	13.693 ± 14.545	–34.872 ± 9.293	–66.070 ± 16.331
q	1.464 ± 0.375	1.511 ± 0.527	–0.208 ± 0.186	–1.753 ± 0.353
chi^2	61,778.2	10,031.9	41.9	169.0

(C) Other Potential Variables

	%H$_2$O in waste	Cl (kg/t) in waste	%O$_2$ at tert duct	PCDF (μg/t) in waste	PCDD (μg/t) in waste	Feed Rate (t/h) waste
a	19.32	0.280	12.8	1,624	13,671	5.0
	16.91	0.132	12.9	408	4,218	4.6
b	15.73	0.140	11.5	770	4,832	7.4
c	42.96	0.137	10.2	435	1,170	6.8
	30.41	0.148	9.2	1,243	8,673	7.0
d	50.33	0.062	8.7	481	2,658	9.5
	45.58	0.111	10.6	327	2,703	9.6
f	14.37	0.177	4.8	626	426	11.3
	33.75	0.241	4.7	771	1,833	13.5
h	8.76	0.067	7.3	89,276*	2,549	6.7
	18.66	0.027	7.3	4,064	36,714*	8.5
i	21.90	0.335	8.6	1,506	10,215	7.1
	9.24	0.322	8.4	263	853	7.3
j	30.87	0.037	7.0	435	4,110	6.2
	36.02	0.077	8.0	272	1,279	8.2

Source: Adapted from Neulicht 1987 and Green et al. 1992.

sensitivities to all variables, unfortunately, the availability of only 15 data sets at the tertiary duct and 19 at the boiler exit limits the possibility of fixing additional parameters.

Concentrating on the variables t, Z, and X, the group obtained poor correlation coefficients when they assumed $Y_i(t)$. These improved dramatically when they assumed $Y_i(t,t^2)$. Adding $Z = CO/100$ as a variable, $Y_i(t,t^2,Z)$ generally improved the correlation coefficients. Adding HCl as a variable led in all cases but one (P at duct) to positive linear coefficients of X in $Y_i(t,t^2,Z,X)$. This indicates that increased HCl generally correlates with increases in chlorinated organic emissions. These parameters are determined fairly sharply as indicated by the (error a_4)/a_4 in several cases. These five parameter equations provide physically reasonable smooth characterizations of the Vicon data for all but the phenol tertiary duct case.

The Danish National Environmental Research Institute has conducted an extensive series of dioxin emission measurements on municipal and hospital incinerators (Manscher 1989b). Measurements were carried out according to a statistical design, which permitted causal interpretation of the correlations found between the dioxin emissions and certain operating parameters. Using multivariate linear (in parameters) regression they found good correlations with

$$\ln Y = a_0 + a_1 \text{RLD} + a_2 O_2 + n \ln \text{HCl} \qquad (1\text{--}6)$$

where Y is the expected total sum of tetra- to octachlorinated dioxins and furans per Nm^3 at 10 percent oxygen, a_0 is a constant for each incinerator, RLD is the relative load deviation from the design load, O_2 is the deviation of oxygen content from 11 percent in the dry flue gas, and $a_1 = 2.53$, $a_2 = 0.23$, and $a_3 = 0.19$. This equation may be reexpressed as

$$Y = a^2(\text{HCl})^n \exp(a_1 \text{RLD} + a_2 O_2) \qquad (1\text{--}7)$$

where $a^2 = \exp a_0$. The exponential dependencies of Y on RLD and O_2 give a large sensitivity to deviations from the design load or optimum combustion conditions. Equation 1–7 is thus physically similar to Eqs. 1–3 and 1–5 in which combustion conditions are expressed explicitly in terms of the combustion variables, absolute temperature, and carbon monoxide.

1.6 THEORETICAL MODELS OF CHC FORMATION

A number of theoretical approaches have been proposed to explain PCDD/PCDF formation processes in incinerators. In brief these include the following:

1. The PCDD/PCDF output represents the unburned PCDD/PCDF in the input feed.
2. Chlorophenols combine in the incinerator to form PCDD/PCDFs.
3. Incompletely combusted hydrocarbon compounds form unchlorinated dioxins or furans; then chlorine donors displace one, two, three, or more hydrogen atoms to form PCDDs/PCDFs.
4. Certain metallic fly ash particles catalyze the combination of unchlorinated dioxins and furans with chlorine from chlorine donors.
5. Reactions between Cl_2 and phenols form PCDDs and PCDFs.

HCl participates indirectly by releasing Cl_2 via the Deacon reaction,

$$2HCl + O \longrightarrow Cl_2 + H_2O \qquad (1-8)$$

Shaub and Tsang (1983) and Penner et al. (1987) have proposed detailed kinetic models of PCDD and PCDF production that focus on detailed congener production. It is thus difficult to correlate their results with the overall chlorinated hydrocarbon yields expressed as total PCDF, PCDD, chlorobenzene (ClBz), or chlorophenols (ClPh) in the previous section. The formation of PCDF and PCDD on fly ash has been proposed by Vogg et al. (1987). Christmann et al. (1989) have found that during combustion and pyrolysis of pure PVC and PVC-cable sheathings in air, PCDD/PCDF are formed in significant amounts up to the ppm range. Thiesen et al. (1989) have shown by studies of real fires and laboratory combustion tests that PVC-containing materials are PCDF/PCDD precursors.

More recently, Gullett et al. (1990) found that PCDDs and PCDFs are mainly formed in a laboratory reactor by reactions between Cl_2 and phenol at rates that are greatest at low temperatures. They suggest that HCl participates indirectly via the Deacon reaction (Eq. 1–8). This reaction is enhanced by the presence of metal chloride catalyst in the ash, most notably $CuCl_2$.

Most recently, Tsang (1990) has shown that most of the binary reaction rates in a 36 × 36 matrix of possible lightweight hydrogen, carbon, and oxygen molecule reactions in a flame are already known. Very few of the reaction rates involving the chlorinated hydrocarbon molecules are known. The large array of reactions that must be considered (~500 to 1,000) suggests that it will be some time before a kinetic approach that considers fluid dynamics and radiation is achieved.

Green et al. (1992) have modified the simple kinetic model previously used in coal-water-gas studies to apply to the waste combustion problem. The model does not include fluid dynamics but uses a simple pyrolysis model and realistic temperature-time histories drawn from experiment. This pyrolysis model was originally developed for a macromolecular model of coal with a unit molecule such as $C_{35}H_{26}O_3A$, which is compatible with coal mass fractions of carbon, hydrogen, oxygen, and ash. The oxygen in the unit molecule is contained in loosely bound CO_2 and H_2O molecules, which are released in the first pyrolysis step. This model is being adapted for waste consisting mainly of plastics, paper, and some foodstuffs. The plastics comprise mostly polystyrene and polyethylene with approximately equal fractions of the monomers $C_2H_3C_6H_5$ and C_2H_4. The garbage is treated as a polymer with loosely bound CO_2 and H_2O molecules and a unit monomer similar to that used in coal but with a much higher hydrogen-to-carbon atom ratio. In the Green et al. model it was assumed that unchlorinated benzene (B_0) is generated from styrene (ST), which is generated from polystyrene (PST) via the reaction

$$PST \longrightarrow ST \longrightarrow C_6H_6 + C_2H_2 \tag{1-9}$$

Unchlorinated phenol (P_0) was generated by the reaction

$$C_6H_6 + OH \longrightarrow C_6H_5OH + H \tag{1-10}$$

Unchlorinated furans (F_0) and dioxins (D_0) were assumed to be produced via

$$2C_6H_5OH \longrightarrow C_{12}H_8O + H_2O + H_2 \tag{1-11}$$

and via

$$2C_6H_5OH \longrightarrow C_{12}H_8O_2 + 2H_2 \tag{1-12}$$

Destructive decomposition paths for B_0 and P_0 mainly to acetylene (C_2H_2) were assumed. When reaction rates were unavailable they were guesstimated so that the major and minor hydrocarbon species that reached the tertiary Vicon duct at 4.7 seconds are in reasonable proportions.

Next Green et al. (1992) pursued the hypothesis that chlorinated organics are trace constituents formed by the chlorination of hydrocarbon products of incomplete combustion. To get into the ballpark of the Vicon data, they assigned single-step chlorination rates intended to mock up the entire series of chlorinated organic congeners. Following the notation of the previous section, they let $B_0 (= Y_{30})$ denote C_6H_6, $B (= Y_3)$ denote the total chlorobenzenes (ClBz), $P_0 (= Y_{40})$ denote C_6H_5OH, $P (= Y_4)$ denote total chlorophenols (ClPh), $F_0 (= Y_{10})$ denote $C_{12}H_8O$, $P (= Y_1)$ denotes total chlorofurans (PCDF), $D_0 (= Y_{20})$ denote $C_{12}H_8O_2$, and $D (= Y_2)$ denote total chlorodioxins (PCDD). They next assumed for each of these cases that chlorination proceeds either via the exothermic reaction

$$Y_{i0} + Cl_2 \longrightarrow Y_i + HCl \qquad (1-13)$$

or the endothermic reaction

$$Y_{i0} + HCl \longrightarrow Y_i + H_2 \qquad (1-14)$$

In both cases it was possible to get into the ballpark of the Vicon tertiary duct results with this simple kinetic model. When the Deacon mechanism and chlorination by Cl_2 is used (Eq. 1-13) the relationships between CHCs and HCl tend to be quadratic. When direct chlorination by HCl (Eq. 1-14) is used, the relationship tends to be linear—all other variables being held constant. Other possible chlorination mechanisms might be considered including the near resonant three-body reactions

$$Y_{i0} + HCL + O \longrightarrow Y_i + H_2O \qquad (1-15)$$

and

$$Y_{i0} + Cl + OH \longrightarrow Y_i + H_2O \qquad (1-16)$$

or the endothermic atomic chlorine reaction

$$Y_{i0} + Cl \longrightarrow Y_i + H \qquad (1-17)$$

Since PVC is the principal source of the HCl, which would be the direct or indirect source of Cl_2 or Cl, one would expect Y_i yields to be monotonically dependent upon PVC input no matter which of Eqs. 1-13 to 1-17 provide the major chlorination mechanism. Figure 1-6, an early example of a kinetic calculation by Green et al. (1992), shows results based upon the

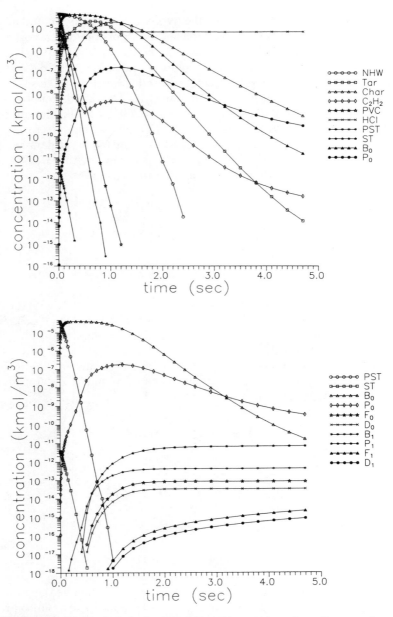

FIGURE 1–6. Concentrations vs. time of species of interest in toxic studies according to a preliminary run of kinetic model of Vicon incinerator emissions. (From Green et al. 1992.)

assumption of Eq. 1–13 and Eq. 1–8 as the chlorination mechanisms. For this model CHC emissions are approximately proportional to (PVC input)^2. For a model based upon Eq. 1–14, CHC emissions are linearly related to PVC.

Work is also currently under way to model the production rates of the various congeners. Perhaps it will be possible to match some observed congener emission factors by fixing relative reaction rates leading to these congeners according to simple physical or chemical rules.

A recent publication of an equilibrium analysis of some chlorinated hydrocarbons in stoichiometric to fuel rich postflame combustion environments (Chang et al. 1991) provides another approximate theoretical model that supports a relationship between CHC production and chlorinated compound input. This work indicates that the potential formation of chlorinated species that are stable at ambient temperatures can arise if locally fuel-rich mixtures penetrate into the lower temperature zones of an incinerator.

1.7 SUMMARY

The discovery and control of fire has served mankind in advancing civilization over many millenia. Very early on cavemen learned by experience that some fuels burn cleanly and others do not. Whenever possible they chose the former for the purposes of cooking, heating, processing clays or metallic ores, and disposing biological material (including corpses to control disease or plague). Thus fuel selection, a precombustion measure, has been a component of clean combustion technology (CCT) since time immemorial.

Good combustion, the second component of CCT, was also discovered very early. Thus good fuel stacking, oven arrangements, gratings, hearths, chimneys to create a draft that provided a good air supply, and manual stoking were also recognized and applied to achieve good combustion.

Postcombustion control, the third component of CCT, in the form of afterburning (as in Dutch ovens), settling chambers, and other methods also have some fairly old roots. However, for the past two decades air pollution control of particulates, acid emissions, and emissions, from hazardous waste incineration have received the greatest attention.

As noted in Section 1.3, in clean coal technology the precombustion component in the form of coal selection and coal cleaning has advanced considerably in the last two decades. However, in the case of municipal waste and particularly medical waste, precombustion measures have received very little consideration to date. In particular, making medical disposables out of materials that burn well and avoiding toxic-producing materials (TPMs) is a lesson yet to be applied. Many of the disposables in the medical waste stream, upon combustion, produce toxic chlorinated hydrocarbon emissions and metals (from plasticizers and pigments). Polyvinyl chloride is a prime example of such material although this issue has been obfuscated by the present day focus on air pollution control devices and by some overly stated conclusions in the literature. In this chapter we have particularly focused on the PVC issue and have assembled studies that contradict the conventional wisdom that there is no correlation between PVC input and CHC output.

Perhaps the most convincing evidence of the efficacy of PVC reduction in lowering PCDD/PCDF emissions are measurements in Hamilton, New Zealand, reported by Bulley (1990). His comparison showed that the Hamilton PCDF and PCDD emissions were much lower than the corresponding emissions at Sutter, St. Agnes, and Cedars Sinai hospitals reported by CARB. To reduce corrosive HCl emissions, New Zealand control authorities have attempted to limit the amount of PVCs used by hospitals. Bulley originally attributed Hamilton's low PCDF/PCDD emissions to the high destruction efficiency achieved by their New Zealand–designed four-stage controlled air incinerator. However, the importance of PVC reduction measures is strongly emphasized in more recent work (Bulley 1991). The correlation of low PCDF/PCDD with low HCl is clearly shown in measurement of waste with low plastic content at Shelly Bay in Wellington (see Chapter 3).

In summary, the experimental, phenomenological, and theoretical components of this chapter all support the hypothesis that chlorinated hydrocarbon input will be reflected in chlorinated hydrocarbon output. The degree of correlation is highly sensitive to combustion conditions as expressed by time, temperature, turbulence, stoichiometry, CO level, H/Cl ratio, and transient changes in combustion condition (i.e., puffs or plugs). Nevertheless, most evidence points to the physically and chemically reasonable rule that reducing the input sources of HCl will generally reduce the output of CHCs. Further supportive evidence will appear in later chapters.

ACKNOWLEDGMENTS

The work of the Clean Combustion Technology Laboratory on toxic emissions was supported, in various parts and ways, by the University of Florida, Tacachale, the Center for Environmental Toxicology Program of the Florida Department of Environmental Regulation, the Tennessee Valley Authority Biofuels Program, and the Florida Energy Office. We thank

Gene Hemp, Donald Price, Gerald Shaffer, Winfred Phillips, Max Jackson, Robert Gaither, Phillip Badger, Richard Strickland, Larry Stokely, and Barry Andrews for various forms of support that made this study possible. We thank Professors Christopher Batich, Kenneth Wagener, Craig Saltiel, David Powell, Jimmy Street, and Charles Schmidt, Bruce Green, and Donald Clauson for their participation during the formulation and data-gathering phases of this study and for their efforts related to our prior publications. Finally, we also thank Xie-Qi Ma, Joseph Blake, Scott Quarmby, Jonathan Carter, Tamara Markl, Christopher Jeselson, Jason Weaver, and Scott Driskell for experimental and computational support.

REFERENCES

Barton, R. G., G. R. Hassel, W. S. Lanier, and W. R. Seeker. "State-of-the-art assessment of medical waste thermal treatment." Report for Risk Reduction Engineering Laboratory, U.S. EPA Contract #68-03-3365 and California Air Resources Board Contract #A832-155 Energy and Environmental Research Corporation, Irvine, CA, 1989.

Blackmore, G., and I. Leibson. 1986. "Clean coal technology." National Coal Council Report, Arlington, VA.

Brown, R. S., K. Pettit, K. J. Mundy, P. W. Jones, and W. Noble. 1989. "Incineration: the British experience." In Proceedings of the Ninth International Symposium, Toronto, Ontario, Sept. 17–22, 1989. Chemosphere 1785–93.

Bulley, M. M. "Medical waste incineration in Australasia." Paper 90-27.5 in Proceedings of the 83rd Annual Meeting of the Air and Waste Management Association, Pittsburgh, PA, June 24-29, 1990.

Bulley, M. M. 1991. Incineration of medical waste: Treating the cause rather than the symptoms. Clean Air (May 1991): 51–53.

Chang, D. P. Y., R. E. Mournighan, and G. L. Huffman. 1991. An equilibrium analysis of some chlorinated hydrocarbons in stoichiometric to fuel-rich post-flame combustion environments. J. Air Waste Manage. Assoc. 41: 947–955.

Christmann, W., D. Kasiske, K. D. Kloppel, H. Partscht, and W. Rotard. 1989. Combustion of polyvinylchloride—an important source for the formation of PCDD/PCDF. Chemosphere 19(1–6): 387–92.

Clement, R., C. Sashiro, G. Hunt, L. LaFleur, and V. Ozvacic. 1989. Chlorinated dioxins and related compounds. In Proceedings of the Ninth International Symposium, Toronto, Chemosphere Sept. 17–22, 1989.

Cross, F. L., H. Hesketh, and P. K. Rykowski. 1990. Infectious waste management. Lancaster, PA: Technomic Pub. Co.

Doucet, L. G. 1991. State-of-the-art hospital and institutional waste incineration: Selection, procurement, and operations. Technical Document Series. Chicago, Il: American Society for Hospital Engineering of the American Hospital Association.

Freeman, H. M. "The United States EPA Pollution Prevention Research Program." Paper 90-41.1 in Proceedings of the 83rd Annual Meeting of the Air and Waste Management Association, Pittsburgh, PA, June 24–29, 1990.

Fry, B., K. Howard, L. Woodhouse, J. Fischer, A. Eli, and M. Goodin. 1990. Technical support document to proposed dioxins and cadmium control measure for medical waste incinerators. Prepared by the Toxic Air Contaminant Control Branch for the California Air Resources Board. Sacramento, CA.

Green, A., ed. 1980. *Coal burning issues.* Gainesville, FL: University Presses of Florida.

Green, A., ed. 1981. *Alternative to oil, burning coal with gas.* Gainesville, FL: University Presses of Florida.

Green, A. 1986. "Clean combustion technology." In Proceedings of Conference *There Is No Away,* University of Florida Law School. Gainesville, FL. Available through EPA-NTIS.

Green, A., and B. Green. 1981. Development of gas-coal combustor. Unpublished, patent No. 4,561,364 issued December 1985: Method of retrofitting an oil-fired boiler to use coal and gas combustion.

Green, A., and K. Pamidimukkala. "Kinetic simulation of the combustion of gas/coal and coal/water mixtures." In Proceedings of the 1st Conference on Combined Combustion of Coal and Gas, Cleveland, OH, 1982.

Green, A., and K. Pamidimukkala. 1984. Synergistic combustion of coal and natural gas. *Energy, the International Journal* 9: 477–84.

Green, A., E. Mize, B. Green, et al. 1985. "Natural gas use to facilitate coal and waste combustion." In Proceedings of the Conference on Select Use of Natural Gas for Environmental Purposes, American Gas Association, Arlington, VA.

Green, A., B. Green, J. Wagner, et al. "Coal-water-gas, an all American fuel for oil boilers." In Proceedings of the 11th International Conference on Slurry Technology, Hilton Head, SC, 1986.

Green, A., G. Prine, B. Green, et al. "Coburning in institutional incinerators." In Proceedings of Air Pollution Control Association International Specialty Conference on Thermal Treatment of Municipal, Industrial, and Hospital Wastes, Air Pollution Control Association, Pittsburgh, PA, 1987.

Green, A., D. Rockwood, J. Wagner, et al. 1988. Co-combustion in community waste to energy systems. *Co-Combustion* ASME FACT. 4: 13–27.

Green, A., C. Batich, D. Powell, et al. "Toxic products from co-combustion of institutional waste. Paper 90-38.4 presented at the 83rd Annual Meeting of the Air and Waste Management Association, Pittsburgh, PA, June 24–29, 1990a.

Green, A., C. Batich, J. Wagner, et al. 1990b. "Advances in uses of modular waste to energy systems." Presented at the Joint Power Generation Conference ASME-IEEE, Oct. 1990 and published in A. Green and W. Lear, eds. 1990. *Advances in solid fuels technologies.* New York: Fuels and Combustion Technology (FACT) Division of American Society of Mechanical Engineers.

Green, A., H. van Ravenswaay, J. Wagner, et al. 1991a. Co-feeding and co-firing biomass with non-hazardous waste and natural gas. *Bioresource Technology* 36:215–21.

Green, A., J. Wagner, and K. Lin. 1991b. Phenomenological models of chlorinated hydrocarbons. *Chemosphere* 22: 1–15.

Green, A., J. Wagner, C. Saltiel, J. Blake, and Xie Qi Ma. "Medical waste incineration with a toxic prevention protocol." Paper 91-33.5 presented at the 84th Annual Meeting of the Air and Waste Management Association, Pittsburgh, PA, June 16–21, 1991c.

Green, A., J. Wagner, C. Saltiel, and M. Jackson. "Pollution prevention and institutional incineration." Presented at American Society of Mechanical Engineers Solid Waste Processing Conference, Detroit, MI, May 17–20, 1992.

Gullett, B. K., K. R. Bruce, and L. O. Beach. 1990. Formation of chlorinated organics during solid waste combustion. *Waste Management and Research* 8: 2203–2214.

Hansen, E. M. 1984. Protocol for the collection and analysis of volatile POHCs using VOST. EPA-600/8-84-007, P384-170042.

Hanson, D. J. 1991. Dioxin toxicity: New studies prompt debate, regulatory action. *Chemical Engineering News* 69(32): 7–14.

Jenkins, A. "Evaluation test on a hospital refuse incinerator at Saint Agnes Medical Center, Fresno, CA." Air Resources Board Test Report No. SS-87-01, 1987.

Jenkins, A. "Evaluation test on a refuse incinerator at Stanford University Environmental Safety Facility, Stanford, CA." California Air Resources Board Test Report No. ML-88-025, 1988a.

Jenkins, A. "Evaluation retest on a hospital refuse incinerator at Sutter General Hospital, Sacramento, CA." CARB Test Report No. C-87-090, 1988b.

Lauber, J. D., and D. A. Drum. "Best controlled technologies for regional biomedical waste incineration." Paper 90-27.2 in Proceedings of the 83rd Annual Meeting of Air and Waste Management Association, Pittsburgh, PA, June 24–29. 1990.

Licis, I., H. Skovronek, and M. Drabkin. 1991. *Industrial pollution prevention opportunities for the 1990s*. EPA/600/S8-91/0-52. Cincinnati, OH: U.S. Environmental Protection Agency.

Manscher, O. H., et al. "The effect of dioxin formation of different chlorine sources in incineration." kd-TEKNIK Miljoproject, Danmark nr.117 and nr.118, 1989a.

Manscher, O. H., et al. "The Danish incinerator dioxin study, Part 1." In Proceedings of the Ninth International Symposium, Toronto, Sept. 17–22, 1989.

McCormack, J. E. "Evaluation test on a small hospital refuse incinerator, St. Bernardine Hospital." Draft Report, CARB Test Report No. C-87-092. San Bernardino, CA, 1988a.

McCormack, J. E. "Evaluation test on a small hospital refuse incinerator, LA County—USC Medical Center. Preliminary Draft, California Air Resources Board Test Report No. C-87-122. Los Angeles, CA, 1988b.

Neulicht, R. "Results of the combustion and emissions research project at the Vicon incinerator facility in Pittsfield Massachusetts." Report 87-16, Midwest Research Institute for the New York State Energy Research and Development Authority, Albany, NY, June 1987.

Penner, S. S., D. F. Weisenhahn, and C. P. Li. 1987. Mass burning of municipal wastes. In *Annual Review of Energy,* Vol. 12, J. M. Hollander, H. Brooks and D. Sternlight, eds. pp. 415–44. Palo Alto, CA: Annual Reviews Inc.

Shaub, W. M., and W. Tsang. 1983. Dioxin formation in incinerators. *Environ. Sci. Technol.* 17: 721–30.

Shaub, W. M., and W. Tsang. 1985. Overview of dioxin formation in gas and solid phases under municipal incinerator conditions. In *Chlorinated Dioxins and Dibenzofurans in the Total Environment II,* L. H. Keith, C. Rappe, and G. Choudhary, eds., pp. 469–87. Stoneham, MA: Butterworth.

Theisen, J., W. Funcke, E. Balfanz, and J. Konig. 1989. Determination of PCDF's and PCDD's in fire accidents and laboratory tests involving PVC-containing materials. *Chemosphere* 19(1–6): 423–8.

Theodore, L. 1990. *Air pollution control and waste incineration for hospitals and other medical facilities.* New York: Van Nostrand Reinhold.

Tsang, W. "The chemistry of hazardous waste incineration." Preprint of paper presented at Eastern Section Meeting of the Combustion Institute, Orlando, Dec., 1990.

Tufts University and Washington Department of Ecology. 1987. *List of toxic and hazardous substances in household waste.* Spokane, WA.

U.S. EPA. 1979. *APTI course 450, Source sampling for particulate pollutants, student workbook.* Research Triangle Park, NC: U.S. Environmental Protection Agency.

U.S. EPA. 1986a. SW-846. *Test methods for evaluating solid waste,* method 0010, 3rd ed. Washington, DC: U.S. Environmental Protection Agency.

U.S. EPA. 1986b. SW-846. *Test methods for evaluating solid waste,* method 0030, 3rd ed. Washington, DC: U.S. Environmental Protection Agency.

U.S. EPA. 1988. 450/3-88-017. *Hospital waste combustion study data gathering phase.* Final Report. Research Triangle Park, NC: U.S. Environmental Protection Agency.

U.S. EPA. 1989. *Method 26—Determination of HCl emissions from stationary sources. Federal Register,* 54(243): 52201. U.S. Environmental Protection Agency.

U.S. EPA. 1990. 625/7-90/009. *Guides to pollution prevention: Selected hospital waste streams.* Cincinnati, OH: U.S. Environmental Protection Agency.

U.S. House of Representatives. 1990. Report 101-952. *The Clean Air Act Amendments.* Washington, DC: U.S. Government Printing Office.

Vogg, H., M. Metzger, and L. Stieglitz. "Recent findings on the formation and decomposition of PCDD/PCDF in solid municipal-waste incineration." Paper presented at the International Solid Wastes and Public Cleansing Association Specialized Seminar on Emission of Trace Organics from Municipal Waste Incineration, Copenhagen, Denmark, Jan. 20–22, 1987.

2

Characterization of Today's Medical Waste

Floyd Hasselriis and Laura Constantine

2.1 INTRODUCTION

Incineration sterilizes and detoxifies medical waste and converts it to innocuous ash, reducing its weight and volume by 90 to 95 percent. However, there is concern about pollutants emitted into the air during the incineration of medical waste, especially acid gases, heavy metals, and dioxins. There is also concern about potentially toxic substances that remain in the ash residues. These pollutants are derived from the waste feed material and generally change in form during the combustion process. The concern increases as hospitals use larger quantities of disposable plastics. In addition, existing incinerator stacks are often short and located close to other buildings.

Although the acid gas and toxic emissions from hospital incinerators represent a relatively small portion of environmental emissions, emissions from hospital incinerators are more likely to have a significant local effect due to their location in urban areas close to residences. An understanding of this medical waste stream and its constituents is the starting point for evaluating methods of pollutant control, including prevention and reduction.

New and existing medical waste incinerators are required to meet increasingly strict emission limits for a variety of pollutants. The new emission limits referred to in Chapter 5 can be met by the installation of air pollution control equipment. However, removing certain materials from the waste stream *prior* to the incineration process could reduce the quantities and types of pollutants produced.

This chapter characterizes medical waste in North America and identifies sources of potentially harmful or toxic substances, so that an evaluation of methods of control can be made.

2.2 DEFINITIONS OF MEDICAL WASTE

Medical waste is defined as all waste generated from health care or health-related facilities. The types of waste classified as medical waste vary according to the institution or department in which they are generated. For example, most of the waste produced by blood banks would include disposable needles and syringes, whereas a research laboratory would produce mostly animal carcasses and organs (Doucet 1987).

The typical types of waste generated by specific areas in a hospital are illustrated in Table 2–1. Medical waste can be classified into two categories: *general waste* and *special waste*. These two waste types are distinct in character and require specific waste treatment and disposal programs because they may contain infectious matter.

General Waste

General waste consists of all waste materials that are not regulated or defined as hazardous, special, or potentially dangerous and do not require special handling and disposal. These wastes are sometimes referred to as nonregulated medical waste (NRMW) (see Chapter 9). The categories of waste defined by the Incinerator Institute of America are useful in describing wastes intended for incineration. General waste, described by types 0, 1, 2, 3, and 4 (Table 2–2), comprises a varying, heterogeneous mixture of paper goods, corrugated cardboard, plastics, food scraps, glassware, metals, and other miscellaneous organics and inorganics.

In hospitals, general waste is produced in all areas, including patient rooms, laboratories, cafeterias, and administrative offices. The typical composition of general medical waste is shown in Table 2–3. This corresponds to an equal mixture of type 0 and type 1 wastes.

Special Wastes

The types of waste classified as special waste require special handling, treatment, and disposal, usually according to specific regulations and guidelines. Such waste may pose potential health, safety, or environmental hazards or may simply be objectionable for disposal with general waste because of appearance or aesthetics.

There are three categories of special waste: *chemical waste, infectious waste,* and *radioactive waste*. These may be in solid, biological, or liquid forms. The typical types of waste contained in each of these categories are listed in Table 2–4.

Most chemical and radioactive wastes are generated in diagnostic and

TABLE 2–1. Hospital Areas Generating Wastes and Their Typical Products

Administration
Paper goods

Obstetrics department including patients' rooms
Soiled dressings; sponges; placentas; waste ampules, including silver nitrate capsules; needles and syringes. Disposables: masks; drapes; sanitary napkins; blood lancets; catheters and colostomy bags; enema units; diapers and underpads; gloves; etc.

Emergency and surgical departments, including patients' rooms
Soiled dressings; sponges; body tissue, including amputations; waste ampules; needles and syringes; Levine tubes; catheters; drainage sets; colostomy bags; underpads; surgical gloves; etc. Disposables: masks; drapes; casts; blood lancets; enema basins.

Laboratory, morgue, pathology, and autopsy rooms
Contaminated glassware, including pipettes, petri dishes, specimen containers and specimen slides; body tissue; organs; bones.

Isolation rooms other than regular patients' rooms
Paper goods containing nasal and sputum discharges; dressings and bandages; leftover food; disposable masks; disposable salt and pepper shakers; etc.

Nursing stations
Ampules; disposable needles and syringes; paper goods.

Service areas
Cartons; crates; packing materials; paper goods; metal cans, drums, etc.; bottles, including food containers, solution bottles and pharmaceutical bottles; wastes from public and patient rooms, including paper goods, flowers etc.; waste food from central and floor kitchens; wastes from x-ray; discarded furniture; rags.

Dust and particulate matter from heating and ventilation equipment; soiled linens and uniforms; lint from washing and drying cycles; empty detergent, bleach, and disinfectant containers.

Teaching and research areas
Paper; bottles; dry rubbish; infectious wastes (mostly animal remains, including carcasses and organs); cadavers and organs from surgery; ashes from crematories.

Food preparation areas
Wooden crates; cardboard and plastic cartons containing food; food trimmings; cans; bottles; aluminum and plastic containers; paper try covers; disposable eating utensils; food wastes; etc.

Source: Oviatt 1968.

research laboratories. These wastes are usually collected, stored, and disposed of separately. Chemical waste may be incinerated with other types of waste, however, radioactive waste is normally shipped to a facility specifically designed to dispose of this material.

Infectious Waste
Infectious waste is sometimes referred to as biomedical, biohazardous, contaminated, regulated, or, as they may be called, red-bag waste, and is

TABLE 2–2. Incinerator Institute of America Waste Types

Waste Type	Moisture (%)	Ash (%)	Heating Value (Btu/lb)
TYPE 0 Trash	10	5	8,500

Highly combustible waste, paper, wood, cardboard cartons, including up to 10% treated papers, plastic, or rubber scraps; commerical and industrial sources.

TYPE 1 Rubbish	25	10	6,500

Combustible waste, paper, cartons, rags, wood scraps, combustible floor sweepings; domestic, commercial, and industrial sources.

TYPE 2 Refuse	50	7	4,300

Rubbish and garbage; residential sources.

TYPE 3 Garbage	70	5	2,500

Animal and vegetable wastes, restaurants, hotels, markets; institutional; commercial and club sources.

TYPE 4 Organic wastes	85	5	1,000

Carcasses, organs, solid organic wastes; hospitals, laboratories, abattoirs, animal pounds and similar sources.

TABLE 2–3. Typical General Waste Characteristics

Waste Stream Components
  Paper and corrugated: 60%
  Plastics: 20%
  Food scraps: 10%
  Metal, glass, inorganic: 7%
  Other miscellaneous: 3%

Waste Stream Properties
  Heating value: 7,500 Btu/lb
  Moisture: 17%
  Incombustible: 8%

usually generated in patient care and laboratory areas of health care facilities. Infectious waste includes materials considered to be potential health hazards because of possible contamination with pathogenic microorganisms. Although this definition seems straightforward, regulatory agencies, health care facilities, and the waste service industry, for example, all have different perspectives, which influence their interpretation of what types of waste should be classified as infectious.

In Chapter 9 infectious waste will be called regulated medical waste (RMW). Distinguishing infectious waste from noninfectious waste is difficult, due to variabilities in disease and in the disease-causing process. The

TABLE 2–4. Special Hospital and Institutional Wastes

Special Waste Classes	Dry and Solid Waste	Biological Waste	Liquid Waste
Radioactive	Paper, plastic, cloth; animal litter, bedding	Animal carcasses, organic tissues	Bulk solvents, LSC vials
Infectious	Pathogenically contaminated paper; plastic and cloth; animal litter, bedding	Injected animal carcasses; organic tissues and body parts	Blood and body fluids
Chemical	Contaminated paper and cloth	Injected animal carcasses	Bulk solvents
Containerized	Natural toxins and food additives; organic chemicals (drugs); animal litter, bedding	Organic tissues	Solvents

following factors are required for disease transmission to occur: a viable, virulent pathogen; a susceptible host; a method for transmitting the pathogen into the host; and exposure to a sufficient quantity of microorganisms to cause infection. Testing all potentially infectious waste for the presence of pathogens is neither warranted nor advocated. Therefore, most states define infectious waste as waste that in all probability contains or could contain pathogenic microorganisms and that, because of the type, concentration, and quantity of these microorganisms, may cause disease in persons exposed to it.

The 10 types of waste from which regulatory definitions for infectious waste are usually drawn are identified in Table 2–5. The composition of typical infectious wastes is shown in Table 2–6. This waste may be described as type 0 waste, except that 1 percent chlorides are specified, to account for chlorinated plastics.

Pathological Waste

Pathological waste, one of the types of infectious waste, closely resembles waste type 4 (see Table 2–2). The principal sources of pathological waste are clinical laboratories and the surgery and obstetrics departments of hospitals. Average pathological waste characteristics are given in Table 2–7. To burn this waste, heat must be applied to drive off the high moisture content. This waste will not sustain combustion, hence supplementary fuel is needed to incinerate it.

TABLE 2–5. Infectious Waste Definitions

Isolation waste

Cultures and stocks of infectious agents and associated biologicals (due to the high
  concentration of pathogens typically present in these materials)

Human blood and blood products

Pathological waste (because of the possibility of unknown infection in the patient)

Contaminated sharps (due to the possibility of undiagnosed disease and the double hazard
  that they present of being able to inflict injury and induce disease)

Contaminated animal carcasses, body parts, and bedding

Waste from surgery and autopsy

Miscellaneous laboratory wastes

Dialysis unit wastes

Contaminated equipment

Source: U.S. EPA 1986.

TABLE 2–6. Typical Infectious Waste Characteristics

Waste Stream Components
  Paper and cloth items: 50–70%
  Plastics: 20–60%
  Glassware: 10–20%
  Fluids: 1–10%

Waste Stream Properties
  Heating value: 8,500 Btu/lb
  Moisture: 8.5%
  Incombustible: 5.0%
  Chlorides: 1.0%

Comparative Composition

Some of the medical waste generated by health care facilities is very
similar to household trash, although the majority is not normally found in
any other waste stream. In order to get a general idea of the comparative
composition of medical waste and municipal solid waste (MSW), some
typical analyses are shown in Table 2–8. The main differences are in the
plastic and paper fractions. The red-bag (infectious) waste contains 41
percent plastic and rubber as compared with 21.4 percent in the clear-bag
(general) waste and 12.8 percent in MSW. MSW contains 54.1 percent
paper and cardboard, compared with 31 percent in red-bag and 39 percent
in clear-bag waste (Brown 1989).

Most infectious waste, excluding pathological waste, is essentially iden-
tical to uncontaminated general waste and therefore resembles a mixture

TABLE 2-7. Typical Pathological Waste Characteristics

Heating value: 1,000 Btu/lb
Moisture: 85%
Incombustible: 5%
Bulk density: 40–60 lb/ft^3

TABLE 2-8. Composition of Red-Bag, Clear-Bag, and Municipal Solid Waste

Material	Red-Bag (Infectious) (%)	Clear-Bag (General) (%)	Municipal Solid Waste (%)
Paper	31.0	36.0	41.9
Cardboard	0.0	3.0	12.2
Plastic	29.0	20.0	11.2
Rubber	12.0	1.4	1.6
Textiles	5.0	2.1	2.9
Food	1.0	11.7	11.9
Yard Waste	0.0	2.0	0.0
Glass	3.2	4.8	7.5
Metals	1.1	7.2	6.0
Fluids	17.7	9.9	0.0
Misc. Organics	0.0	1.9	6.4
Total:	100.0	100.0	100.0

Source: Adapted from Brown 1989.

of the Incinerator Institute of America's Waste types 0, 1, 2, and 3 (see Table 2–2). However, the typical breakdown of waste stream components differs substantially when general waste (see Table 2–3) is compared to infectious waste (see Table 2–6).

2.3 WASTE GENERATION

Infectious waste is generated by a wide variety of health care facilities. Some of the typical types of institutions that produce infectious waste include blood banks, clinics, dentists' offices, funeral homes, hospitals, laboratories, nursing homes, physicians' offices, and veterinary hospitals. The total number of generators in each of the above categories, as reported in the literature on a national and regional basis, and the quantity of infectious waste produced by these generators are presented in the following sections.

Approximately 465,000 tons of infectious waste are generated in the

United States each year by 377,000 health care facilities. The quantity of infectious waste generated nationally, by facility type, is illustrated in Table 2–9. There is a high variability in the amount of infectious waste produced by the different health care facilities, as illustrated in this table. Hospitals, which comprise only 2 percent of the total number of generators, produce the greatest quantity (approximately 77 percent) of infectious waste among the different types of institutions (U.S. EPA 1990a,b).

Variability also exists in the quantity of infectious waste produced by facilities within each generator type. While much of this variability can be attributed to differences in facility size, types of services, and specialty offered, part of it is a result of differences between state regulatory definitions and requirements and internal waste management policies and practices at each of the individual institutions. At some institutions, because of the difficulties and/or uncertainties in managing waste segregation programs, substantial quantities of general waste are inadvertently intermixed with infectious waste. In such cases, the quantity of waste treated as infectious waste is often greater than that actually classified as infectious. At other institutions, policies and protocols established by infection control committees and similar environmental safety groups broaden the definition of potentially infectious waste to include more sources and materials than are covered by applicable regulations and guidelines. For example, many hospitals across the country categorize all patient-contact waste as potentially infectious. Under this designation, as much as 60 to 90 percent of all hospital waste must be managed and disposed of as potentially infectious.

TABLE 2–9. Sources and Quantities of Infectious Waste Generated Nationally

Type of Generator	Number of Generators	Total (t/yr)	Generated Per Facility (lb/month)
Hospitals	7,100	359,000	8,400
Nursing homes	12,700	29,600	390
Physicians' offices	180,000	26,400	24
Clinics	16,700	16,700	180
Laboratories	4,300	15,400	600
Dentists offices	98,400	7,600	13
Veterinarians	38,000	4,600	20
Funeral homes	20,400	3,900	32
Blood banks	900	2,400	440
Total	377,300	465,600	

Pounds per year per Capita: 4.2

Source: U.S. EPA 1990a.

This practice is usually justified as being consistent with Centers for Disease Control (CDC) Guidelines of August 1987 entitled "Recommendations for Prevention of Human Immunodeficiency Virus (HIV) Transmission in Health-Care Settings." These guidelines, which are commonly referred to as *Universal Precautions*, basically recommend that "all patients be considered potentially infected with HIV and/or other blood-borne pathogens." However, according to recent CDC interpretations and clarifications, waste from patient rooms under the Universal Precautions system is not considered infectious unless otherwise specifically identified as infectious. Thus, the data provided in Table 2–9 are an estimate of the quantities of infectious waste generated per month for an average facility in each generator type (U.S. EPA 1986, 1990a).

2.4 PLASTICS AND METALS CONTENT

The plastics content of medical waste has increased significantly over the past 10 years. During the late 1970s, only 10 percent of the medical waste stream was plastics. By the late 1980s, however, plastics comprised more than 30 percent. This increase can be attributed to the expanded use of disposable plastics by the health care industry. Disposable plastics have replaced many of the glass items and textiles used for clothing.

The amount of polyvinyl chloride (PVC) plastic found in an incinerator's waste feed material is closely related to the production of hydrogen chloride, and may influence dioxin and furan emissions. Health care facilities can reduce the quantity of PVC plastics disposed of either by recycling these products or by replacing the items made out of this material with items made from alternate materials. Providers of these products are moving rapidly in this direction.

The amount of heavy metals contained in an incinerator's waste feed material directly determines the quantity found in the emissions. The obvious sources of metals found in medical waste include needles, surgical blades, and foil wrappers. Less obvious are the fillers, stabilizers, colorants, and inks that are used in the production of plastics. Medical wastes, which contain a large amount of transparent and colorless plastics, are likely to have lower metal contents than municipal waste, which contains printed matter and colored and pigmented items.

2.5 PLASTIC DISPOSABLES

Increasingly, health care facilities are using disposable items. A study carried out in Ontario hospitals is useful to illustrate the quantities and nature of these items.

TABLE 2–10. Hospital Waste Generation Rates by Major Plastic Type

Type of Plastic	Size of Hospital			
	Small (kg/bed/day)	Large (kg/bed/day)	Small (g/patient/day)	Large (g/patient/day)
Latex	0.002	0.005	0.007	0.017
Polyethylene	0.140	0.073	0.500	0.248
Polypropylene	0.043	0.008	0.155	0.003
Polystyrene	0.027	0.009	0.074	0.031
Polyurethane	0.006	0	0.023	0
Polyvinyl chloride	0.057	0.065	0.202	0.222

Source: Ontario Ministry 1988.

TABLE 2–11. PVC Conent of Disposable Plastics

Type of Hospital	Percent PVC by weight	Vinyl gloves (%)	IV sets (%)	Syringes (%)	PVC (g/patient/day)
Medium general	13.5	42			0.20
Regional cancer center	11.7	30	20	25	0.28
Pediatric hospital	7.5			40	0.08
Large general	10.6				0.22

The annual quantities of disposable plastic items were obtained from the hospitals' purchasing and inventory records (see Tables 2–10 and 2–11). The materials of construction and weight of each item were found from suppliers and distributors. The composition was determined by both olfactory and visual flame tests of the items.

Flame tests provide a qualitative indication when the results of burning a small piece of the sample in a propane burner flame are observed. A PVC plastic gives a yellow-green color and an acrid choking odor. Other plastics characteristically produce sooty flames, have self-extinguishing characteristics, tend to melt, or produce characteristic odors (Ontario Ministry 1988).

Six basic types of plastic are contained in medical waste: polypropylene, polyvinyl chloride (PVC), polystyrene, polyethylene, polycarbonate, and mixed plastics. The percentage that each type contributes to the total plastic composition of medical waste is illustrated in Table 2–10.

Commercially available PVC resins used to make disposable medical items are generally one of the following:

- Flexible items with a high plasticizer content: 36% PVC
- Rigid items with a minimum of plasticizer: 56–64% PVC

The major PVC items are as follows:

- Vinyl gloves
- Intravenous (IV) administration sets
- Syringes and needles

The percentage of total plastics and the distribution varied with the hospital (Table 2–12). Cafeteria items included cutlery, plates, and cups. Other nonmedical items such as toothbrushes, combs, and identification cards were investigated. Data on the various plastic types are shown in Table 2–13.

2.6 METALS CONTENT OF WASTE MATERIALS

Paper and plastic products contain significant amounts of trace metals. Many materials used as fillers, such as clay, contain trace metals. Analysis of samples of components of MSW gives an indication of the actual quantities normally present in the basic components of MSW. Many of these components are also present in medical waste. However, many of these components would be present in noninfectious general wastes, which may be landfilled rather than burned in a medical waste incinerator.

Typical metals contents of some common components of municipal waste are presented in Chapters 4 and 5. The various paper products have

TABLE 2–12. General Waste Data

Material	Heat Value (Btu/lb)	Density (lb/ft^3)	Ash (%)	Moisture (%)
Polyethylene	20,000	40–60	0	0
Animal fats	17,000	50–60	0	0
Polyurethane foam	13,000	2	0	0
Milk cartons	11,330	5	1	4
Coffee grounds	10,000	25–30	2	20
Rags	8,600	10–15	2	5
Newspaper	8,000	7	1.5	6
Wood	8,000	12–20	3	10
Brown paper	7,250	7	1	6
Corrugated paper	7,040	7	5	5
Magazines	5,250	35–50	23	5
Food waste	1,800	25–30	6	75

TABLE 2–13. Characterization of Hospital Waste

Material	Moisture (%)	Heat Value (Btu/lb)	Bulk Density (lb/ft^3)
Plastics	0–1	13,900–20,000	50–75
Plastics, PVC, syringes	0–1	9,600–20,000	5–144
Alcohol, disinfectants	0–0.2	11,000–14,000	48–62
Swabs, absorbents, gauze, pads, swabs, garments, paper, cellulose,	0–30	5,600–12,000	5–62
Beddings, shavings, paper, fecal matter	10–50	4,000–8,100	20–45
Human anatomical	70–90	800–3,600	50–75
Animal infected anatomical	60–90	900–6,400	30–80
Fluids, residuals	80–100	0–2,000	62–63
Glass, sharps, needles	0–1	0	450–500
Glass	0	0	3,000

similar levels of lead, cadmium, chromium, and nickel. The PET and HPDE bottles have relatively high lead contents. Nickel is high for expanded polystyrene (EPS) and the other plastics. The emissions of lead and cadmium from medical waste incinerators are much lower than those from the municipal waste incinerators (Hasselriis 1990).

2.7 HEATING VALUE OF MEDICAL WASTES

The heat liberated by combustion of medical wastes, expressed as Btu per pound, is an important factor. Incinerators have a built-in combustion air supply, which directly limits their burning capacity. Therefore, the number of pounds per hour that can be burned depends directly on the heating value per pound. The emissions from an incinerator can also be put on a pound basis, to obtain the pounds per hour of each pollutant. The heating values of the components of general waste and medical waste are listed in Tables 2–12 and 2–13. There is a wide range in heating value from 5,000 to 20,000 Btu per pound. The density listed affects the number of pounds of waste that can be loaded into a given incinerator charging hopper volume.

The heating value of medical waste is affected substantially by the heating values of its components, especially plastics. Heating values can be calculated by the procedures shown below.

Polyethylene Plastics

Polyethylene (C_2H_4) is one of the most common plastics. Its total molecular weight is $2 \times 12 + 4 \times 1 = 28$. Its carbon weight fraction is 24/28, or

85.7 percent, and its hydrogen is 4/28, or 14.3 percent. The heating value of pure polyethylene is approximately

$$(85.7/100) \times 14,100 + (14.3/100) \times 61,000 = 20,807 \text{ Btu/lb}$$

Organics Containing Oxygen

Organic materials such as wood, paper, vegetable matter, meat, and alcohols contain oxygen in addition to hydrogen and carbon. The following combustion equation applies:

$$C_xH_yO_z + O_2 \longrightarrow H_2O + CO_2 + \text{Heat}$$

Newspaper contains 50.02 percent carbon, 6.21 percent hydrogen, and 43.77 percent oxygen, by weight. The oxygen already combined with the hydrogen does not produce heat, hence the oxygen must be subtracted from the hydrogen with which it combines. Oxygen has an atomic weight of 16, and the hydrogen in water is 2, so 16/2, or 8 units, of oxygen combine with the hydrogen.

The volume equation is found by dividing the weight fractions by the molecular weights. For instance, the volume fraction of carbon is based on $50.02/12 = 4.17$ percent carbon:

$$C_{(4.17)}H_{(6.21)}O_{(2.74)} + (4.35)O_2 \longrightarrow (3.1)H_2O + (4.17)CO_2$$

The *weight* equation, per 100# of paper, is found by multiplying the volumes by the molecular weights:

$$C_{(50\#)}H_{(6.2\#)}O_{(43.8\#)} + (139\#)O_2 \longrightarrow (55.8\#)H_2O + (183.5\#)CO_2$$

The heating value per pound of dry newspaper is

Carbon:	0.50 lb	$\times$ 14,100 Btu/lb =	7,050 Btu/lb
Hydrogen:	0.062 lb	$\times$ 61,000 Btu/lb =	3,788 Btu/lb
Oxygen:	$-(0.438/8)$ lb	$\times$ 61,000 Btu/lb =	$-3,337$ Btu/lb
Total Heating Value:			7,500 Btu/lb

The Dulong formula (Domalsky 1986) may be used to estimate the higher heating value (HHV) of combustibles, given the weight percent of the elements:

$$HHV\ (Btu/lb) = 145.5[\%C] + 609.6([\%H] - [\%O]/8) + 40[\%S] + 10[\%N]$$

Organics Containing Chlorine

Polyvinyl chloride (PVC) contains chlorine, which reacts with some of the hydrogen to form hydrochloric acid, HCl. The remainder of the hydrogen is free to form water. The reaction is

$$PVC + Oxygen \longrightarrow Water + Carbon\ Dioxide + Hydrochloric\ Acid$$

The PVC volume equation is

$$C_2H_3Cl + (5/2)O_2 \longrightarrow (1)H_2O + (2)CO_2 + (1)HCl$$

The weight equation is

$$(62\#)C_2H_3Cl + (80\#)O_2 \longrightarrow (18\#)H_2O + (88\#)CO_2 + (36\#)HCl$$
$$(24\# + 3\# + 35\#) + (80\#) \longrightarrow (142\#)$$

The heating value of PVC can be estimated from the oxidation of carbon and hydrogen with approximate allowance for the formation of the stable compound HCl (Green, private communication).

Carbon:	$[24\#/(24+3+35)\#] \times 14,100$ Btu/lb =	5,458 Btu/lb
Hydrogen:	$[2\#/(24+3+35)\#] \times 61,000$ Btu/lb =	1,968 Btu/lb
HCl:	$[36\#/(24+3+35)\#] \times 1,000$ Btu/lb =	581 Btu/lb
Total Heating Value:		8,007 Btu/lb

Pure PVC has roughly the same heating value as paper, but about half that of polyethylene. The chlorine content of pure PVC is 35.45/62.45 = 56.76 percent. The fillers and plasticizers that are normally used reduce this percentage.

Polyethylene plastic and oil have similar heating values, whereas paper and PVC have only about 40 percent as much heat content. Although various combustibles require different amounts of oxygen per pound of fuel to complete combustion, they require about the same amount of oxygen per unit of heat released. Hence the amount (by weight) of waste that can be burned with a given amount of oxygen depends upon its heating value.

2.8 SUMMARY

Medical wastes include a wide range of components, of which a large percentage are disposable items consisting of plastics of various types. A fraction of these contains a large percentage of chlorine or fluorine, and hence greatly influences the halogen content of the mixture. The content of halogenated plastics can be substantially reduced by identifying these items and removing them or substituting them with nonhalogenated materials.

Heavy metals found in plastics and paper are derived from pigments, fillers, and inks. The content of toxic metals such as lead, cadmium, chromium, and mercury in paper and plastics could be reduced by elimination or substitution with nontoxic compounds. The halogen and metals contents of products purchased by hospitals could be controlled by specification of toxic-free products, obliging vendors to make appropriate substitutions.

REFERENCES

Brown, H. L. Thomas Jefferson University Hospital waste characterization study. Drexel University. 1989.

CDC. "Guideline for handwashing and hospital environmental control, 1985." Centers for Disease Control, U.S. Department of Health and Human Services, Washington, DC. NTIS PB85-923404, 1985.

CDC. "Recommendations for prevention of Human Immunodeficiency Virus (HIV) transmission in health-care settings." Centers for Disease Control, U.S. Department of Health and Human Services, Washington, DC, *Morbidity and Mortality Weekly Report,* Vol. 36, August 21, 1987.

Domalski, E. S., T. Jobe, and T. A. Milne. 1986. *Thermodynamic data for biomass conversion and waste incineration.* SERI/SP-271-2839, Boulder, CO: Solar Energy Research Institute.

Doucet, L. G. "Institutional waste incineration problems and solutions." Paper presented at Incineration of Low Level and Mixed Wastes Conference, St. Charles, IL, April 1987.

Hasselriis, F. "Relationship between waste composition and environmental impact." Presented at the 83rd Annual Meeting of the Air and Waste Management Association, Pittsburgh, June 24–29, 1990.

Jack Faucett Associates. 1989. Draft, "Final infectious and chemotherapeutic waste plan."

Morrison, R. 1987. *Hospital waste combustion study—Data gathering phase.* Radian Corp., Research Triangle Park, U.S. EPA Contract 68-02-4330.

Ontario Ministry of the Environment. A study of the quantity and type of disposable plastics used by different hospitals in Ontario. Ontario Ministry of the Environment, Waste Management Branch, No. 148907, March 1988.

Oviatt, V. R. 1968. Status report—Disposal of solid wastes. *Hospitals* 42: December 16.

U.S. EPA. 1986. EPA/530-SW-86-014, PB86-199130. *EPA guide for infectious waste management*. Washington, DC: U.S. Environmental Protection Agency, Office of Solid Waste.

U.S. EPA. 1987. EPA/530-SW-87-0216. *Municipal waste combustion study: Emission data base*. Washington, DC: U.S. Environmental Protection Agency, Office of Solid Waste.

U.S. EPA. 1989. Federal Register, 40 CFR Parts 22 and 259. Washington, DC: U.S. Environmental Protection Agency, Office of Solid Waste.

U.S. EPA. 1990a. EPA/530-SW-90-051A, *Medical waste management in the United States—First interim report to congress*. Washington, DC: U.S. Environmental Protection Agency, Office of Solid Waste.

U.S. EPA. 1990b. EPA/530-SW-90-087A, PB91-130187. *Medical waste management in the United States—Second interim report to Congress*. Washington, DC: U.S. Environmental Protection Agency, Office of Solid Waste.

3

Medical Waste Management in Australasia

Michael M. Bulley

3.1 INTRODUCTION

Prior to 1984 the incineration of medical waste in New Zealand had a very poor reputation, and most of this type of waste was sent to landfill sites for disposal. This poor reputation was a result of the changes brought about by the introduction of sterile disposable plastics, which initially tended to be largely made of polyvinyl chloride (PVC). The great increase in the calorific value of the wastes as a result of the introduction of the throwaway plastics overwhelmed the existing incinerators, and they responded by emitting billowing clouds of acidic black smoke. It was found that some control on smoke emissions could be achieved by the careful charging and firing of the waste, but this was never enough to achieve the invisible stack plume that the public and authorities demanded.

At Auckland Hospital the poor level of control achieved was considered so inadequate that the use of the incinerator was banned by the authorities. At other hospitals attempts to commission poorly designed scrubbers did little to solve the problem of acid smut and smoke emissions. The net result of the exercise was that virtually all waste was sent to landfill, and those incinerators that continued to operate did so preferentially at night.

In 1984 several factors resulted in a review of the neglected medical waste disposal policies. The initiators of this impetus were as follows:

1. A realization that not only were most of the existing landfill sites poorly designed in terms of leachate control, but, more importantly, space was rapidly running out.
2. The growing concern expressed by hospital infectious control staff about the advisability of landfilling infected or potentially infected

wastes. This concern was boosted by the public concerns about AIDS and to a lesser extent hepatitis B.
3. The growing opposition by landfill operator unions to the handling of medical wastes at the landfill sites.
4. Press reports about dogs scavenging among poorly landfilled medical waste and of children finding used hypodermic needles at landfill and other sites.

3.2 THE REVISED INCINERATION POLICY

The approach to medical waste disposal in Australasia was somewhat different than that in the United States. Reasons for this difference included the inherited concept of "best practicable means" as a control philosophy, the local absence of a strong development-oriented pollution control equipment supply industry, the comparatively small size of the industry, and a relatively unpolluted environment.

In 1985 a new policy on incineration was introduced in New Zealand, which was followed in turn by Victoria and New South Wales in Australia. The objectives of the policy were as follows:

1. To ensure the complete burnoff of any smoke or particulates generated in the primary combustion chamber (i.e., to maintain an invisible stack throughout the incineration cycle).
2. To destroy as far as possible, by means of afterburning, any unavoidable products of incomplete combustion that would be generated in the primary combustion zone.
3. To destroy any cytotoxic materials that may be charged into the incinerator.
4. To ensure that the incinerator operation was as operator friendly as possible.

To achieve the above objectives, all new incinerators were required to have a minimum of two combustion chambers. The final afterburning chamber was required to simultaneously operate at (1) a temperature in excess of 1,000°C (1,832°F) at all times, (2) an excess of oxygen of 6 to 10 percent, and (3) a residence or retention time of all off-gases of at least 1 second. The states of New South Wales and Victoria in Australia required in turn a minimum afterburner operating temperature of 1,100°C (2,012°F).

The selection of the 1,000°C temperature benchmark was based largely on theoretical work done by Cassitto (1985), which showed that a 99.99 percent reduction of 2,3,7,8-tetrachloro-dibenzo-p-dioxin (2,3,7,8-TCDD) could be achieved at the following temperatures and residence times: 6

years, 477°C; 46.5 minutes, 727°C; 1 second, 977°C; 0.5 second, 1,000°C; and 4 milliseconds, 1,227°C. These theoretical calculations compared favorably with actual experimental work done by the U.S. Environmental Protection Agency (EPA).

The authorities have recognized that one of the emissions from a medical waste incinerator will be hydrogen chloride, but they do not yet automatically require a scrubber to be installed for the control of these emissions. To this end the limits on allowable emissions vary between 100 and 400 mg/Nm³, with the understanding that exceeding these limits would require the installation of a tail gas treatment plant for the following reasons:

1. There is an existing procurement program to minimize the amount of PVC plastic used by hospitals. As this progresses there is the expectation that stack HCl levels will reduce to acceptable levels.
2. The installation of a wet scrubber, while reducing HCl levels by around 90 percent, will give a cold wet plume with no inherent plume rise, and such a plume can result in unwanted high ground level peaks of HCl.
3. At relatively low levels of emission, the substantial additional cost of a scrubbing system, even with a reheat facility, would result in a minimal or negative ground or lung level benefit.
4. Such a control system, because of the inherent corrosion problems associated with the presence of hydrogen chloride together with sulfurous/sulfuric acid, would need to have either exotic materials of construction or a comprehensive system of temperature controls, and hence would be disproportionately expensive.

This approach can be best illustrated by Figure 3–1, which shows the predicted ground level concentrations that would result from various levels of hydrogen chloride at a temperature of 700°C from a 15-meter high stack versus the cold emission of 20 mg/Nm³ from a scrubber with no plume rise. Stable conditions and a 2 m/s wind speed are used.

3.3 OPERATIONAL EXPERIENCE

The first incinerator installed in New Zealand that was supposed to meet the then new 1985 conditions was an imported unit. As soon as the unit arrived, it was evident that the afterburning chamber was totally inadequate. The purchasers tried to obtain help from the equipment suppliers without success. To meet the required standards, the entire afterburning system had to be redesigned and modified by a local engineering company. Operating experience obtained during the transition stage confirmed that

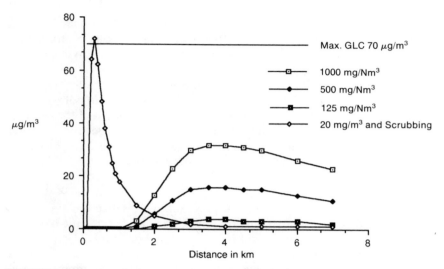

FIGURE 3–1. Predicted HCl ground level concentration: stable conditions, wind speed 2 m/s.

the larger afterburning chamber and the resulting increased afterburner retention time resulted in the improved ability to operate smokelessly and also offered a greater measure of operator friendliness.

The second incinerator to be installed was a three-chamber controlled-air unit, wholly designed and manufactured in New Zealand. This unit was installed in Hamilton, New Zealand, and has a general configuration as shown in Figure 3–2.

Dioxin destruction is considered representative of the likely fate of all products of incomplete combustion. The choice of dioxins as an indicator is made on the grounds of their known resistance to thermal destruction and the recognition that dioxin emissions have a major impact on environmental health risks (Von Burg 1988).

It has taken some time to quantify the actual dioxin emissions from different incinerators due to analytical costs and the fact that until very recently only one laboratory in Australasia was capable of doing the analyses. To assess the validity of the design concept. the measured dioxin concentrations from three different incinerators have been compared.

The Wellington Hospital Incinerator

This is a two-stage pyrolytic incinerator (Abbott 1990a) with a maximum afterburner temperature during testing of 950°C and an overall afterburner

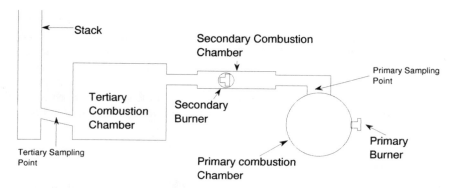

FIGURE 3–2. General configuration of the Hamilton incinerator.

residence time of <0.1 second. This incinerator is representative of an older style of incinerator and has a history of poor and smoky operation. For the purposes of the test the waste was selected and hand-loaded into the primary combustion chamber. During the test runs the unit operated without visible emissions. The purpose of conducting the test in this manner was to simulate what could be considered near-optimum performance. The results were taken as indicative of the upper performance capability of the older incinerators. It is worth noting in passing that this unit has now been permanently decommissioned.

The Incinerator at Hamilton

This unit has been discussed elsewhere (Abbott 1989) and is essentially a three-stage unit, with a tracer gas measured afterburner residency time of 1.5 seconds and an afterburner temperature of >1,050°C. This unit was taken to represent the performance of the new type of incinerators. Results obtained from a comparable design unit in Sydney, Australia, were found to be similar.

An Incinerator at Shelly Bay in Wellington

This unit (Abbott 1990b) incinerates harbor and airport quarantine waste mixed with medical waste. During the test the material incinerated was roughly a 50/50 mix of the two wastes. As the quarantine wastes have a low plastic content, the results can be considered representative of what could happen if PVC plastics were reduced.

The dioxin results obtained in terms of 2,3,7,8-chlorinated furans and dioxins are shown in Figures 3–3 and 3–4. The results of the total measured tetra, penta, hexa, hepta, and octa congeners of the furans and dioxins are shown in Figures 3–5 and 3–6. Both sets of graphs show a consistent difference between the three incinerators selected. This would indicate that the move toward high afterburner residence times and temperatures has been in the right direction. There is also an indication that the removal of PVC plastics will result in lower dioxin levels as well as reduced hydrogen chloride emissions.

3.4 EMISSION TESTING AT HAMILTON

The initial testing program was conducted at the Hamilton plant during March and May of 1989, with samples of the flue gas extracted immediately

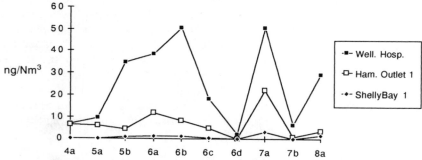

4a=2378TCDF;5a=12378PeCDF;5b=23478PeCDF;6a=123478HxCDF;6b=123678HxCDF;6c=123789HxCDF;6d=234678
HxCDF;7a=1234678HpCDF;7b=1234789HpCDF;8a=OCDF

FIGURE 3–3. 2,3,7,8-furan congeners, ng/Nm^3 at 10 percent O$_2$.

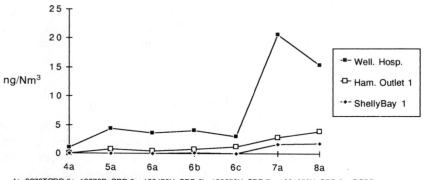

4a=2378TCDD;5a=12378PeCDD;6a=123478HxCDD;6b=123678HxCDD;7a=1234678HpCDD;8a=OCDD

FIGURE 3–4. 2,3,7,8-dioxin congeners, ng/Nm^3 at 10 percent O$_2$.

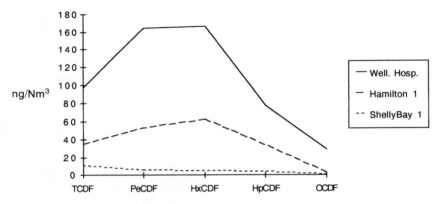

FIGURE 3–5. Furan congeners, ng/Nm³ at 10 percent O₂.

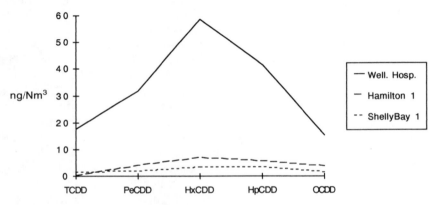

FIGURE 3–6. Dioxin congeners, ng/Nm³ at 10 percent O₂.

downstream of the primary and tertiary combustion chambers. Test techniques developed for these tests included the following:

1. The use of fused silica glass sampling probes to withstand the high temperatures, particularly from the tertiary chamber.
2. The development of a Xenon 127 gamma isotope tracer measurement technique to determine the true tertiary residence time. This method was developed by the Institute of Nuclear Studies of the Department of Scientific and Industrial Research (DSIR), Wellington.
3. The use of a Nordic modification of the U.S. EPA Method 5 sampling train to sample PCDDs and PCDFs.

Standard measuring techniques and methods were used to monitor the other parameters.

The carbon monoxide, oxygen, and temperature levels recorded at the outlets of the primary and tertiary chambers for the two tests are shown in Figure 3–7. These readings show the different combustion regimes in the two chambers. Incomplete combustion in the primary chamber was characterized by high CO levels (2 to 5 percent), and in the tertiary chamber complete combustion was characterized by low CO (<1 ppm) and high temperature.

The tertiary chamber carbon monoxide readings were consistently below 1 ppm except for an 8-minute excursion in Test 1, which resulted from a temporary imbalance caused by an incorrect attempt to change the excess oxygen levels. During this period a maximum reading of 160 ppm CO was recorded, which also resulted in a visible darkening on the sample train filter and a visible light smoke emission from the stack for about half a minute.

The hydrogen chloride emission tests showed the stack emission to average 100 mg/Nm^3, dry, at 10 percent oxygen, with a range between 62 and 160 mg. This average emission level equals the plant license limit. The gas residence times within the tertiary chamber as tested by the Xenon 127 isotope method showed the residence times to be 1.7, 1.7, and 1.4 seconds. The stack particulate emission level was found to be in the range of 50 to 60 mg/Nm^3, dry, at 10 percent oxygen. This particulate loading was not unexpected, as a characteristic of the incinerator is for it to operate with an invisible plume throughout the total burn cycle.

The measured values of the various congeners of PCDD and PCDF at the outlets of the primary and tertiary chambers are shown in Figures 3–8 and 3–9. The values are expressed as ng/Nm^3, dry, and at 12 percent CO$_2$. The primary chamber values are shown as columns relating to the values on the left-hand axis, and the tertiary chamber values as lines relating to the values on the right-hand axis. As can be seen, the PCDD and PCDF concentrations are 100 to 1,000 times more concentrated in the primary chamber. No traces of 2,3,7,8-TCDD were detected at the exit of the tertiary chamber in either of the two tests. The most likely explanation for the high dioxin levels measured at the exit of the primary chamber was that these and other products of incomplete combustion are actually formed in the primary chamber rather than being present in the waste from the start.

Figures 3–8 and 3–9 show the individual group patterns in each chamber, with the more highly chlorinated congeners tending to predominate in the tertiary chamber. The tests show that PCDDs are more readily destroyed than PCDFs; this may be more apparent than real as the cause could be the additional formation of PCDFs by the breakdown of PCDDs.

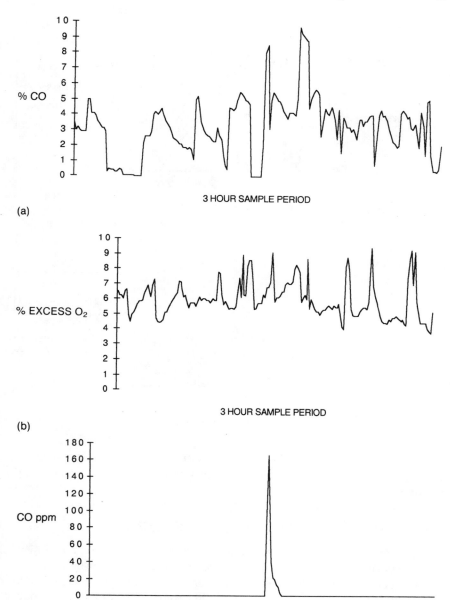

FIGURE 3–7. (a) Primary chamber CO, (b) tertiary chamber excess oxygen, and (c) tertiary chamber CO, ppm.

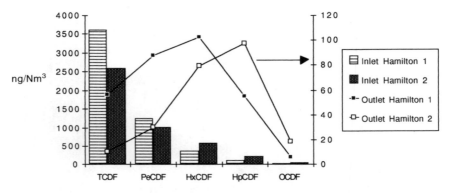

FIGURE 3–8. Furans, ng/Nm^3 at 12 percent CO_2.

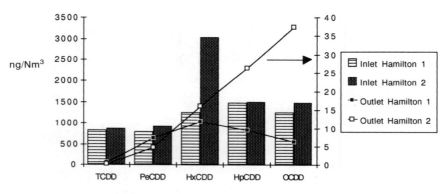

FIGURE 3–9. Dioxins, ng/Nm^3 at 12 percent CO_2.

When the 2,3,7,8-TCDD equivalent toxicities of the two tests were assessed, it was found that the PCDFs contributed to more than 80 percent of the final values.

3.5 COMPARISON WITH OTHER RESULTS

No references have been found showing dioxin results before and after an afterburning chamber for an incinerator dedicated to burning clinical wastes. There are some results of dioxin concentrations from hospital waste incinerators at Sutter Hospital, St. Agnes Hospital, and Cedars Sinai Medical Center (CARB 1987a,b,c) published by the California Air Resources Board, which could be used for comparative purposes. Figures 3–10 and 3–11 compare the results from these hospitals with those obtained at Hamilton for the basic groupings of PCDD and PCDF. As can be seen, the Hamilton test result is considerably lower.

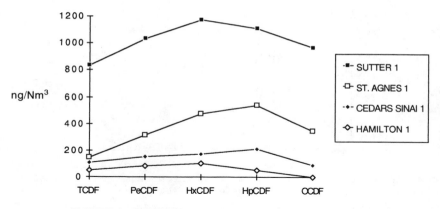

FIGURE 3–10. PCDF comparison, ng/Nm³ at 12 percent CO_2.

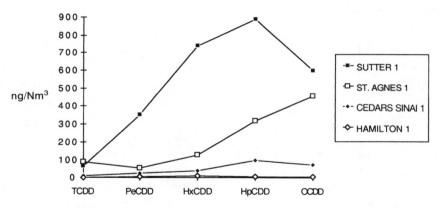

FIGURE 3–11. PCDD comparison, ng/Nm³ at 12 percent CO_2.

Figures 3–12 and 3–13 compare those congeners normally used in calculating 2,3,7,8-TCDD toxic equivalents. Again the Hamilton test results are lower. There is a common pattern in the concentration of individual compounds, particularly for the furans. It is possible that residence time and, to a lesser extent, afterburner temperatures could be responsible for the differences in concentration of the compounds in the emissions from the various hospitals.

3.6 ENVIRONMENTAL IMPACTS OF INCINERATORS

Recent health risk assessment work on the risks associated with emissions from municipal waste incinerators (Levin et al. 1991) indicates that the

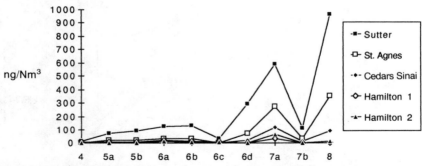

4 = 2378 TCDF; 5a = 12378PeCDF; 5b = 23478PeCDF;6a = 123478HxCDF; 6b = 123789HxCDF; 6c = 123789HxCDF;
6d = 234678HxCDF; 7a = 1234678HpCDF; 7b = 1234789HpCDF; 8 = OCDF

FIGURE 3–12. Furan comparison, ng/Nm^3 at 12 percent CO_2.

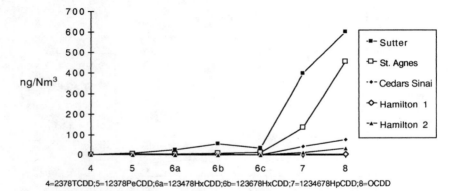

4=2378TCDD;5=12378PeCDD;6a=123478HxCDD;6b=123678HxCDD;7=1234678HpCDD;8=OCDD

FIGURE 3–13. Dioxin comparison, ng/Nm^3 at 12 percent CO_2.

following emissions (in order of decreasing importance) have an impact upon the health risk of the surrounding community:

1. Toxic equivalents of dioxins and furans (TEDFs)
2. Cadmium
3. Polynuclear aromatic hydrocarbons (PAHs)
4. Polychlorinated biphenyls (PCBs)
5. Arsenic
6. Chrome
7. Beryllium

The emissions from medical waste incinerators are not unlike those from municipal waste incinerators. On this basis, the above tabulation could, with the addition of hydrogen chloride, be used as a point of departure in assessing the emission control techniques that could be used for medical waste incinerators.

The above emissions can be divided into two broad categories. The first is those emissions generated to a greater or smaller extent within the incinerator itself, such as TEDFs, PAHs, and PCBs, which are products of incomplete combustion. The second is those emissions that result entirely from pollutants or their precursors being introduced into the incinerator, such as cadmium, arsenic, chrome, beryllium, and hydrogen chloride.

Products such as TEDFs, PAHs, and PCBs, which are generated in the incinerator, arise as products of incomplete combustion within the primary combustion chamber. The higher the calorific value of the waste being incinerated, the greater the amount of local oxygen depletion that will occur during incineration within the primary chamber. This in turn results in an increase in the amount of products of incomplete combustion generated.

Plastics, which have calorific values similar to liquid fuels, now commonly make up 40 to 50 percent of medical waste by volume. Given this high proportion of high calorific material in medical waste, it is inevitable that products of incomplete combustion are generated within the primary combustion chamber, even if it was initially designed to operate under excess air conditions. To destroy this unwanted byproduct, a modern incinerator will include at least one specially designed afterburner. These afterburners are required by control authorities to retain all primary chamber off-gases at a minimum temperature of 1,100°C (2,012°F) for a period of at least 1 second, under turbulent conditions and in the presence of 6 to 10 percent excess oxygen. These conditions are considered the minimum required for the adequate thermal destruction of these products of incomplete combustion.

Plastics are one of the major sources of introduced pollutants. The introduction of pollutants is achieved in several ways, including the use of PVC plastics. The chlorine content of PVC (roughly 45 percent by weight) is converted during combustion into either hydrogen chloride gas or a mixture of hydrogen chloride and chlorine gases (chlorine being generated if no water vapor is present), together with trace amounts of dioxins and furans, and other chlorinated products of incomplete combustion.

PVC was one of the first disposable plastics to be used within the hospital system. Early experiences of the acid corrosion that arose when PVC plastics were incinerated started the conversion to nonchlorinated plastics such as polyethylene, polystyrene, and polycarbonates. This change was not complete, as it was found that most alternative plastics were not suitable for storing or conveying blood. To this day items such as blood bags, blood tubing, tubing connectors, and valves are still largely supplied in PVC plastics. Alternative plastics that are suitable for the safe storage and transfer of blood and blood products are now available. There is, however, a reluctance within the medical fraternity to use these alternative plastics because to do so would be to deviate from what is known as normal standard practice. Such a deviation could, if the medical procedures resulted in a poor outcome, become a reason for possible litigation.

Most commercial plastics contain additives to give them a desired flexibility or rigidity, resistance to ultraviolet light, and fire resistance. Common fillers, stabilizers, and pigments used in the plastics industry include compounds of barium, cadmium, zinc, calcium, magnesium, phosphates, strontium, tin, antimony, lead, chrome, selenium, cobalt, molybdenum, nickel, iron, and titanium (Hasselriis 1990). For example, colors are often made as follows:

White: titanium dioxide, zinc sulfide, zinc oxide
Red/red orange: cadmium, chrome, molybdenum, nickel, iron, lead
Orange/yellow/buff: cobalt, chrome
Pearl: lead carbonate

The yellow pigments used to give the required yellow color to the plastic bags and buckets have a ≥ 90 percent chance of having cadmium chrome as a color base. An alternative nontoxic organic pigment is available at a comparable price. Manufacturers, however, prefer the cadmium chrome pigment because it is not temperature sensitive and hence can tolerate a wider range of extrusion temperatures and processing parameters than the organic pigment.

Plastics designated specifically for medical use would not be expected to contain toxic additives. This expectation should, however, be verified before it is accepted as fact. Plastics supplied for nonmedical use either by the hospital or a waste contractor should be far more suspect. The reason for this is that the order runs are rarely large enough to warrant a specific run formulation. In most cases the formulation tends to be the one used for a similar bulk run. As a consequence, properties in terms of rigidity, ultraviolet light, and fire resistance are automatically passed on, without any real need for some of those properties. In this way heavy metals can quite easily end up in plastic materials destined for incineration. The red or yellow bags and buckets used in the disposal of infectious wastes generally contain cadmium and/or lead salts.

3.7 EMISSION CONTROL

There are two ways of controlling the emissions of hydrogen chloride and heavy metals. The first, and unfortunately the most common, method is to treat the symptom and install a collection device on the incinerator outlet so as to remove the pollutants prior to the incinerator off-gas being vented to atmosphere. Such devices are normally either a combination of a fabric filter followed by a wet scrubbing unit or a series of high efficiency wet scrubbers. Both of these configurations can be very efficient in removing the unwanted pollutants from the off-gas. Unfortunately, the pollutants have simply been transferred from one medium to another, and further steps are often required to ensure that the pollutant can finally be disposed of in an environmentally acceptable manner.

The second and only reliable method for removing these components from the incinerator effluent streams, whether in the stack off-gases, the ash and collected particulates, or the scrubber effluent water, is to ensure their removal prior to their entering the system. This policy of substitution not only treats the cause of the problem, but it also makes sense in that the unwanted pollutants never enter the waste stream. It has an additional benefit, in that workers in the upstream supply industry are not exposed to the contamination risks of handling the various toxic materials in bulk.

The elimination of these components can be achieved by good procurement practices and effective waste separation at the source. There is no technical reason why plastics used in the health care system should not be substantially halogen- and heavy metal–free. The concept of changing a feed stock is not new. What is relatively new is the need to persuade all the different interest groups and suppliers of the reasons for requiring such a change and the benefits it will bring. This is an area that has only recently been highlighted, and in Australasia, formulation queries to some plastics suppliers meet with vague answers, simply because many of the materials and pigments used in the plastics industry are traded under brand names, where the actual chemical composition is kept intentionally vague.

Hospital procurement staff have a growing awareness of the situation, which has been brought to their attention by recently published standards and legislative guidelines (EPAV 1989; SANZ 1990; SPCC-NSW 1990), as well as in-house policies and growing pressures from environmental groups. The Victorian guidelines also suggest that hospitals review their procedures with a view toward minimizing the use of disposable items that can be used only once. Central purchasing offices already give preference whenever possible to purchasing items made of biodegradable materials when these items are likely to be disposed of in a landfill. There is every reason to believe that this policy will be adopted for items to be disposed of by incineration, provided substitutes are available at a comparative cost.

Intravenous bags of various kinds are a major large volume source of PVC plastics disposed of by incineration. The main reason for this is that until fairly recently PVC was the only plastic material that could be used for storing blood. At the moment there is only one manufacturing source in Australasia, and it has a virtual market monopoly. This plant is a U.S.-owned subsidiary and will, in time, provided there is sufficient domestic pressure on the U.S. parent company, change the plastic formulation. Experience with similar problems indicates that unless a considerable amount of local pressure is exerted on the Australian subsidiary, the changeover is more likely to be later, rather than sooner.

3.8 FEASIBILITY OF SUBSTITUTION

Substitute plastics for PVC plastics with acceptable blood storage and handling characteristics are now available (Green et al. 1990). These substitute products include the following:

Blood bags and tubing: ethylene vinyl acetate copolymers are now available as Elvax 3120 from Dupont; Kraton from Shell; and from Horizon polyers in the United States.
Tubing connectors and valves: polycarbonates, polyacrylates, polystyrene, and various other copolymers are now available.

Because the waste removal industry often purchases bags and containers from the general supply, such material could be contaminated. Plastics supplied to the export food industries use nontoxic pigments and fillers. The waste removal industry can specifiy and obtain plastics compatible to those required by the export food industry. Steps along these lines will help reduce heavy metal emissions from incinerators. It is hoped that a continued program of replacing toxic materials with nontoxic substitues will produce a continued improvement across the board.

Plastics supplied directly to hospitals or by waste contracting companies, which collect and contain medical wastes that are required to be incinerated, seldom have a known detailed formulation. Plastics supply companies, like paint companies, regard actual formulation details as trade secrets and are reluctant to give any details. Smaller supply companies very often simply do not know the true chemical composition of the formulations they are using, and in fact may be actively discouraged from making detailed enquiries.

A possible way around this problem is to insist that the plastics meet the standards required by the food industry for products being exported to Japan and the United States. The food export industry has had some embarrassing moments overseas related to toxic materials in plastics, and the food export industry has come down hard on local plastic supply companies.

Virtually all plastics available today are based on ethylene as a feed stock. These plastics are not biodegradable. It is, however, worth mentioning that totally biodegradable plastics known as PHBV (poly-3-hydroxybutyrate-3-hydroxyvalerate) plastics are now commercially available (Anonymous 1989). Polyhydroxybutyrate is obtained from the fermentation of glucose (or any naturally occurring carbohydrate) and is mixed in varying proportions with hydroxyvaleric acid to give various grades of copolymers (Anonymous 1990). PHBV plastic starts biodegrading upon contact with microorganisms. It is totally renewable, nonhazardous, and biocompatible and can be used as surgical pins, sutures, bone replacements and plates, blood vessel replacements, and biodegradable implants for the long-term administration of drugs.

3.9 ALTERNATIVES TO INCINERATION

Incineration is still somewhat controversial, particularly among environmental groups. The major argument against incineration appears to be about the amount of carbon dioxide liberated into the atmosphere and the need to minimize this greenhouse gas. The secondary argument is about the emissions of heavy metals and products of incomplete combustion, such as dioxins. Some concern has also been expressed about the fact that incineration discourages recycling, even though few if any generally acceptable recycling programs exist for medical waste. Alternative strategies for waste that is not clearly of an anatomical nature propose that the waste be handled in one of the following ways:

1. Be landfilled
2. Be treated in an autoclave and then landfilled
3. Be treated in a purpose-built microwaving unit and then landfilled
4. Be treated in a chlorinated hammermill, with the liquid effluent being discharged into a sewer and the solids at a landfill site

Most people accept that direct disposal into a landfill site is unacceptable, but some still feel that the other options should be used. The three remaining options, however, have problems of their own.

Sterilization

There is no guarantee that the material is adequately sterilized in any of the three processes. In autoclaving, live steam has to contact the bio-organism for a discrete period of time to be effective. The greater the bioburden, the longer the contact time needed, and autoclaving times in excess of one hour at steam temperatures of 135°C have been shown to be required. There is also no guarantee that the steam will come into contact with every bit of waste charged into the autoclave. In theory, the sterilization process can be monitored by seeding the waste with known cultures. In practice, the test results only become available long after the waste has been disposed of.

The microwaving process guarantees only that the waste will be held at 96°C for a maximum of 30 minutes. These conditions are not adequate to kill all known bioorganisms. The chlorinated hammermill method offers the most reliable sterilizing method, provided the bleach used is both fresh and of adequate strength.

Leachate Problems

None of the three sterilizing methods mentioned above will alter the toxicity of any chemical that may be present in the waste. In addition, any soluble material will leach from the waste either during the treatment process or at the landfill site. Early tests show that the condensate discharge from an autoclave can be so contaminated that the effluent can be classified as a hazardous waste in its own right, bearing out these concerns. The addition of chlorine to a slurry of medical waste is more likely to increase the overall chemical toxicity than decrease it.

Reaction Within the Landfill

Over and above any leachate problems, all the biodegradable material in the waste will decompose under anaerobic conditions into primarily methane and carbon dioxide, together with minor amounts of a whole host of other products of decay, in concentrations levels of milligrams or micrograms per cubic meter. This cocktail of chemicals will be emitted as seepage gases from the landfill for 20 to 30 years at a rate of around 3 cubic meters per day for every 10 metric tonnes of waste (HMSO 1990). Thus, alternative processes that are considered by many to avoid emissions to atmosphere do in fact add pollutants to the atmosphere, and often in a more dangerous form. These landfill gases have two additional disadvantages: (1) they are liberated at ambient conditions at ground level, and (2) their overall toxicity will be higher, as these gases have not been subjected to the high level of thermal destruction under oxidizing conditions that are inherent in the design of a modern incinerator.

In this day and age acronyms have become very popular, and terms such as NIMBY (not in my backyard), BANANA (ban any new activity near anything), and PIISEBY (put it in someone else's backyard) have become popular buzzwords expressing conservative options. Perhaps it is time for industry and control authorities to use a RESCUE (rational environmental strategies countering unreasonable extremism) approach.

3.10 SUMMARY

Installations in Australia and New Zealand that have afterburning temperatures in excess of 1,100°C (2,012°F), with residence times at those temperatures of more than 1.5 sec and an excess oxygen content of 5 to 10 percent have been shown to be viable and operate at low particulate and dioxin emission levels. The substitution of PVC plastics and elimination of heavy metals as pigments and fillers will allow such units to meet strict emission limits without the need for additional tailgas treatment.

REFERENCES

Abbott, N., et al. 1989. Hospital waste disposal by incineration. Conference Institute of Waste Management (N.Z. Inc.), Wellington, New Zealand.

Abbot, N. 1990a. Flue gas measurements from a waste incinerator at Wellington Hospital. NECAL Service Report S90/522C, May 18, 1990.

Abbott, N. 1990b. Flue gas measurements from a waste incinerator at Shelly Bay, Wellington. NECAL Service Report S90/523C, August 28–29, 1990.

Anonymous. 1989. *Biocycle* (March): 58–60.

Anonymous. 1990. *Waste Manage. Today (News J.)* 3(17): 8.

CARB. 1987a. Evaluation retest on a hospital refuse incinerator at Sutter Hospital, Sacramento, CA. California Air Resources Board Test Report C-87-090. Sacramento, CA.

CARB. 1987b. Evaluation test on a hospital refuse incinerator at St. Agnes Medical Center, Fresno, CA. California Air Resources Board Test Report ARB/SS-87-01. Sacramento, CA.

CARB. 1987c. Evaluation test on a hospital refuse incinerator at Cedar Sinai Medical Center, Los Angeles, CA. California Air Resources Board Test Report ARB/SS-87-11. Sacramento, CA.

Cassitto, L. Destruction of the organo-chlorinated micropollutants in combustion processes. 7th Miami International Conference on Alternative Energy Sources, 1985.

EPAV. Manual for the management and disposal of biomedical wastes. Publication No. 268, Environmental Protection Authority, Victoria, 1989.

Green, A., et al. "Toxic products from co-combustion of institutional waste." Paper in the Proceedings of the 83rd Annual Meeting of the Air and Waste Management Association, Pittsburgh, June 24–29, 1990.

Hasselriis, F. "Relationship between waste composition and environmental impact." Paper in the Proceedings of the 83rd Annual Meeting of the Air and Waste Management Association, Pittsburgh, June 24–29, 1990.

HMSO, 1990. Landfilling wastes. U.K. Department of the Environment Waste Management, Paper No. 26. London.

Levin, A., et al. 1991. Comparative analysis of health risk assessments for municipal waste combustors.'' *J. Air Waste Manage. Assoc.* 41(1): 20–31.

Manual for the management and disposal of biomedical wastes. Publication No. 268, Environment Protection Authority, Victoria, 1989.

SANZ. New Zealand Standard 4304: Health care waste management. Standards Association of New Zealand, Wellington, 1990,

SPCC-NSW. Guidelines for pathological biomedical waste incinerators. State Pollution Control Commission New South Wales, 1990.

Von Burg, R. 1988. Toxicology update: TCDD. *J. Appl. Toxicol.* 8(2): 145–48.

4

Avoidance of Metals in Biomedical Waste Incinerators

Daniel P. Y. Chang, Harold Glasser,
and Capt. Donald C. Hickman

4.1 INTRODUCTION

Interest in biomedical waste incinerators (BMWIs or simply MWIs) has grown because of highly publicized problems associated with disposal of products generated from hospital and laboratory operations and the observation that incinerator emissions often contain relatively high levels of toxic metals, e.g., cadmium (Cd), hexavalent chromium (Cr^{+6}), lead (Pb), mercury (Hg), polychlorinated dibenzo-p-dioxins (PCDDs), and polychlorinated dibenzofurans (PCDFs). Furthermore, many MWIs constructed before 1990 have not been equipped with air pollution control (APC) equipment for particulate matter, have relatively poor stack design (especially in California), and are located in close proximity to populations.

This chapter discusses both known and suspected impacts resulting from the introduction into the MWI stream of materials that contain metals. Also discussed are possible benefits associated with the removal of these materials from waste streams. A brief review of the literature is provided for the reader to gauge the relative importance of metals emissions from MWIs, municipal solid waste incinerators (MSWIs), and hazardous waste incinerators (HWIs). It is followed by an analysis of the data obtained from a series of tests of eight MWIs in California by the Air Resources Board (ARB) from August 1986 to April 1989 (Jenkins 1987a,b, 1988a,b, 1989; McCormack 1988a,b,c). The analysis focuses upon possible relationships among the metals and the PCDDs and PCDFs and examines the efficacy of APC equipment that was in place at the time of testing.

A possible link between metals and elevated PCDD and PCDF levels is mentioned to further highlight the potential importance of reduction of chlorine and metal inputs in the waste stream. Toxic metals in many

73

cases have dominated health risk assessments. Complete elimination of all metals or chlorine is not considered feasible, nevertheless we suggest that hospital administrators not only stop purchasing products containing chlorine and certain metals, but also improve waste segregation practices in order to reduce the major problems. Such a strategy for chlorinated products has been applied by others (Bulley 1990; Green et al. 1990a) and could possibly eliminate the need for costly additional APC equipment.

4.2 BACKGROUND

Biomedical waste (MW) is characteristically heterogeneous and can be broken down into three categories: infectious waste, chemical waste, and radioactive waste. Chemical wastes are regulated under the Resource Conservation and Recovery Act (RCRA). Large quantities are not incinerated in MWIs, though small quantities are known to be present in the MW stream (Drum 1990). Radioactive wastes are regulated by the Nuclear Regulatory Commission (NRC) and are not purposely introduced into MWIs, though they may be present at low levels in tissues. Infectious waste is classified as red-bag waste by most facilities in the United States. *Red-bag* refers to the red plastic disposal bags used to designate infectious waste items. Infectious waste can consist of (1) human and animal anatomical wastes, (2) disposable equipment, instruments, utensils, and like items from a patient who is known or suspected to have a communicable disease, (3) laboratory wastes such as pathological specimens, body fluids, excreta, and disposable fomites, and (4) emergency and operating room specimens and disposable fomites. Infectious waste may also contain toxic components such as waste pharmaceuticals and cytotoxic agents that are exempt from RCRA (Lee, Huffman, and Shearer, 1988). A regulatory definition of infectious waste has not been promulgated yet, though regulated medical waste is defined as waste that has been listed in 40 CFR 259.30(a) (Lee and Huffman 1991).

General wastes, such as from offices, kitchens, and so forth, have been mixed with MW for economic considerations, i.e., reduced disposal costs and when heat recovery has been an ancillary goal of incineration. Infectious waste accounts for approximately 15 percent of the total waste stream in facilities that incinerate both biomedical and general waste (Doyle, Drum, and Lauber 1985). Table 4–1 provides an indication of typical contents of the mixture. However, it is our opinion that the practice of mixing general waste should be avoided because it promotes overcharging the incinerator and disregard for beneficial segregation. Incineration of infectious waste is an appropriate use of MWI to reduce both handling and

TABLE 4-1. Typical Cross Section of Mixed Biomedical Waste

Artificial linens	Paper
Flowers	Waste food
Cans	Diapers
Plastic cups	Syringes
Scalpels	Tweezers
Rubber gloves	Pathological objects
Blood test tubes	Petri dishes
Test tubes	Dropper bottles
Medicine bottles	Drop infusion equipment
Transfusion equipment	Suction catheters
Bladder catheters	Urinal catheters
Colostomy bags	Hypodermic needles
IV tubing	Packaging material

Source: Adapted from Morrison 1987.

liability issues, though segregation of disposable metallic objects such as surgical instruments and needles should be given serious consideration for reasons to be described later.[1] Ease of use and economic pressures have resulted in a widespread dependence on disposable instruments by the health care industry. Sharps containers are disposed of in red-bag waste and typically contain hypodermic needles, razors, and disposable scalpels.

Polyvinyl chloride, (PVC) is a common polymer used for the bags, sharps containers, and intravenous (IV) tubes. Marrack (1988) found that in two hospitals in Houston, Texas, PVC accounted for 9.4 percent of the total weight of the red-bag waste, and total plastics accounted for 14.2 percent. Combined biomedical and general waste typically contains 20 percent plastics, but levels as high as 30 percent have been observed (Doyle, Drum, and Lauber 1985). In comparison, municipal solid waste usually contains 3 to 7 percent plastics. The composition of MW was also determined by an ARB survey in California. A breakdown by category is given in Figure 4-1 of the waste delivered to on-site and regional incinerators. Note that the larger proportion of type 4 wastes received by regional facilities is indicative of the increased cost of disposal at regional facilities.

The U.S. Environmental Protection Agency (1988) estimated that about 3.2×10^6 tons of biomedical waste are generated each year, although the actual figure is not known precisely. According to a 1983 American Health Association survey, about 70 percent of the U.S. MW stream is incinerated

[1] We are not aware of any evidence that biomedical waste has a greater potential for infection than domestic waste or conversely that domestic waste has a lesser potential for infection than biomedical waste!

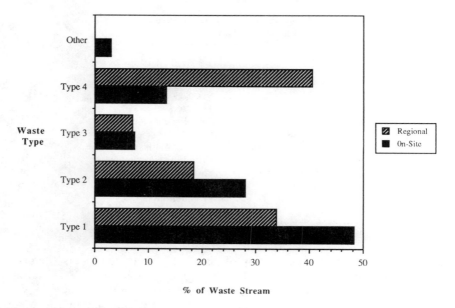

Brief synopsis of waste types:

    Type 1--Cardboard, wood, paper, and miscellaneous wastes
    Type 2--Plastics and rubber
    Type 3--Garbage, food wastes, cans, and bottles
    Type 4--Pathological wastes, human and animal remains

FIGURE 4-1. Percent of each type of waste incinerated for on-site and regional facilities. On-site data are based on 106 survey responses, which represented 61 percent of the reported mass flow. Regional data are based upon 3 responses, which represented 100 percent of the reported mass flow. (From Woodhouse 1989.)

on-site, with one half of the remaining 30 percent receiving pretreatment before being landfilled, and the remainder incinerated at regional facilities (Lee, Huffman, and Shearer 1988). If these figures hold, about 7.4×10^3 t/d of MW were incinerated in comparison with 78.4×10^3 t/d of municipal solid waste (MSW) in 1990 (Anonymous 1990). By comparison, one concludes that *MW currently represents roughly 10 percent, on a mass basis, of the solid waste incinerated in this country.* This figure reflects a roughly fourfold increase in MSW incineration since 1985. The significant fact here is that new MSW incineration facilities were required to meet stringent control standards, while existing MWI facilities have remained virtually uncontrolled.

    Tables 4-2 and 4-3 compare metal concentrations in fly ash and total PCDD and PCDF concentrations from a variety of sources. Additional

TABLE 4–2. Average Cadmium and Lead
Concentrations from Medical Waste
Incinerators, Municipal Solid Waste
Incinerators, and Hazardous
Waste Incinerators

Facility	Concentration in Fly Ash	
	Cadmium (μg/g)	Lead (μg/g)
MWI[a]		
B	1,790	18,100
C	850	12,600
D	666	8,990
E	211	2,640
F	738	24,200
G	504	6,480
H	279	21,800
MSWI[b]		
A	42	4,000
B	185	—
HWI[c]		
A	890	85,500
B	140	3,100
C	1,120	25,600
D	4,000	98,000

[a] Air Resources Board (ARB) series of tests, see Jenkins (1987–1989) and McCormack (1988a,b,c).
[b] Greenburg, Zoller, and Gordon 1978.
[c] Oppelt 1987.

data regarding comparisons of metal concentrations in the flue gas can be found in Barton and Seeker (1990). It is worth noting that metals concentrations *within* fly ash appear to be comparable to or higher than reported values for MSW, but lower than for HWI, whereas the total metals concentration of MWI appears to be lower than for most MSWI if the MSWI did not employ a particulate control system. At the same time, PCDD and PCDF concentrations on the fly ash appear to be lower than MWI for older units, but much greater than that of new units with good combustion, acid gas, and particulate control.

The presence of substantially higher concentrations of chlorine in MWI may lead to mobilization of some metals that would otherwise exhibit lower volatility. Experimental evidence from HWI (Carroll and Waterland 1990) and theoretical calculations for MWI indicate that this can occur (Barton and Seeker 1990). Furthermore, it is currently unknown whether

TABLE 4–3. Average Polychlorinated Dibenzodioxin (PCDD) and Polychlorinated Dibenzofuran (PCDF) Concentrations from Medical Waste Incinerators, Municipal Solid Waste Incinerators, and Hazardous Waste Incinerators

Facility	Concentration in Fly Ash	
	PCDD (ng/m^3)	PCDF (ng/m^3)
BMWI[a]		
B	91	328
C	—Not Available—	
D	339	511
E	31	68.8
F	31	145
G	349	1,240
H	545	1,460
Royal Jubilee	142	40.7
MSWI		
A[b]	4,400	9,400
B[b]	3,550	3,490
C[b]	2,800	10,560
D[b]	2,040	2,480
Marion County, OR[c]	1.5	0.6
Neustadt, Germany[c]	80	95
Albany, NY[c]	300	88
Montreal, Quebec[c]	0.01	0.02
HWI[d]		
A	16	56
B	ND	7.3
C	ND	3.0
D	1.1	ND

[ND = not detected.]
[a] See Tables 2–5 and 2–6.
[b] Beychok 1987.
[c] Huffman 1987.
[d] Oppelt 1987.

the presence of metallic pigments distributed throughout plastics results in increased partitioning to the gas phase than would occur with another matrix.

4.3 ANALYSIS OF SOURCE TEST DATA

The CARB test program provided a unique data set from which health risks could be assessed and possible relationships among waste inputs, operations variables, equipment type, and emissions could be sought. It should be borne in mind that these source tests were not part of a research program where parameters were carefully controlled, but instead were obtained under typical operating conditions. The likelihood of finding strong correlations among the variables and emissions from a test data set involving eight different combustion devices of different manufacture was considered small. In some cases, the samples from a given facility were not even drawn at the same time, though generally within about a week. Nevertheless, such a data collection presented a rare opportunity. An effort to glean whatever information might be present was warranted. Tempered with information gained from the test reports and other studies, some insights might be obtained.

Information regarding the types of incinerators tested, the *indicated* temperatures, and the feed rates is shown in Table 4-4. Six of the incinerators tested were of the dual-chamber, controlled-air variety, while two were single-chamber units with afterburners. Six of the incinerators were located at hospitals, five of those being equipped for heat recovery so that general wastes were fed to the incinerator. Based upon discussions with knowledgeable sources, kitchen wastes were likely segregated and not incinerated (Johnson 1990). Facility A was a regional incinerator burning primarily "infectious wastes." Facility F was located at a university medical center where a higher proportion of laboratory wastes might have been present. In the test report for facility E, it was noted that there was insufficient "infectious" waste to sustain operation for the duration of the test burn, so general wastes were fed to the unit. These are significant pieces of information since the proportion of waste types might have differed substantially at the latter facilities.

In all facilities except facility H, the average CO concentration was less than 100 ppm. Facility H had an average CO concentration of greater than 500 ppm. Facility E had a particularly low average CO concentration, at less than 10 ppm. Concentrations of particulate matter and metals, measured before the APC equipment, if present, and corrected to 12 percent CO_2, are shown in Table 4-5. In some tests, total particulate matter was measured, while in others PM_{10} was measured. For comparability a *very rough* correction was applied to yield both values for every facility based upon optical size distribution data from one facility. The PCDD and PCDF concentrations and their 2,3,7,8-tetrachloro-dibenzo-*p*-dioxin equivalents[2] (TCDDEQV) are given in Table 4-6. Measurements made downstream of

---

[2] California Department of Health Services methodology for 2,3,7,8-tetrachloro-dibenzo-*p*-dioxin equivalents (TCDDEQV).

TABLE 4–4. Combustion Device Summaries from CARB Source Tests[a]

Facility	Combustor	Manufacturer and Model	Average Feed Rate (lb/h)	(h/day)	(t/y)[c]
A	Two-chamber (primary at 1,600–1,800°F and secondary at 1,800–2,200°F) (only burn "medical waste")[b]	Thermtec (#800)	675	24	1,900 1,600[d]
B	Two-chamber (primary at 1,600–1,800°F and secondary at 1,800–2,000°F) with waste heat recovery system	Ecolaire 1500	980	16	1,850
C	Two-chamber with heat recovery	Thermtec EP800AR	551.2	4	260
D	Two-chamber (primary at 1,500–1,600°F and secondary at 1,800–2,000°F) with waste heat recovery system	Environcontrol	783	7	650 1,500[d]
E	Single-chamber with afterburner (stack at 1,400°F) (only burn infectious waste)	Burney-The Burner J81DS, J813	31.5	9	34 55[d]
F	Two-chamber (primary at 1,700–2,000°F and secondary at 1,900–2,100°F)[e]	Ecolaire (custom)	675	8	640 900[d]
G	Two-chamber (primary at 1,300°F and secondary at 1,600–2,300°F) with waste heat recovery	Thermtec 7-SA	474	9	500
H	Single-chamber (1,150°F, with an afterburner at 1,400°F)	MCI Energy (custom)	30.5	8	29

[a] From individual source test reports, unless otherwise noted.
[b] Ramsey 1990; California Air Resources Board 1989a.
[c] California Air Resources Board 1989a. Based upon a five day work week, 47 weeks per year and the (lb/h) and (h/day) figures reported, unless otherwise noted.
[d] Woodhouse 1989.
[e] University medical research facility, which may have a different waste composition.

TABLE 4-5. Average Emission Concentrations for Particulates and Metals, Corrected to 12% CO_2 (Before Controls)[a]

Facility	Particulates (g/dscm)	Corrected PM$_{10}$ (g/dscm)[b]	As (μg/dscm)	Cd (μg/dscm)	Cr (μg/dscm)	Cr^{+6} [c] (μg/dscm)	Fe (μg/dscm)	Pb (μg/dscm)	Mn (μg/dscm)	Hg (μg/dscm)	Ni (μg/dscm)	Total Metals (less Hg) (μg/dscm)
A[b]	0.48	0.26	8.23E + 00	2.28E + 02	1.79E + 02	1.44E + 01	3.25E + 03	1.46E + 04	1.94E + 02	4.78E + 02	6.32E + 01	1.85E + 04
B	0.14[d]	0.14	1.87E + 01	4.00E + 02	1.78E + 01	NM	5.01E + 02	3.91E + 03	3.62E + 01	NM	ND	4.88E + 03
C	0.46	0.25	5.34E + 00	3.31E + 02	2.60E + 01	3.00E + 00	7.40E + 02	5.67E + 03	5.28E + 01	1.91E + 00[e]	2.15E + 01	6.85E + 03
D	0.52[d]	0.52	1.25E + 01	3.43E + 02	4.99E + 01	NM	8.09E + 02	4.67E + 03	3.12E + 01	NM	2.71E + 01	5.95E + 03
E	0.15	0.08	7.20E − 01	3.23E + 01	1.83E + 01[e]	3.26E − 03[e]	7.60E + 01	4.04E + 02	4.71E + 00[e]	5.10E − 01[e]	5.49E + 00[e]	5.42E + 02
F	0.16[d]	0.16	ND	1.69E + 02	2.79E + 01	1.95E + 01	3.09E + 02	3.89E + 03	2.01E + 01	NM	1.79E + 01	4.46E + 03
G	0.21[d]	0.21	5.83E − 01[e]	9.48E + 01	1.63E + 01	NM	7.32E + 02	1.26E + 03	1.96E + 01	NM	1.69E + 01	2.14E + 03
H	0.57	0.31	1.35E + 00	1.59E + 02	1.01E + 01	2.75E + 00	1.24E + 03	5.01E + 03	1.24E + 01	1.23E + 04	1.22E + 01[e]	6.45E + 03

ND = not detected; NM = not measured.

[a] From individual source test reports.

[b] Corrected PM$_{10}$ values were calculated based upon a factor of 0.55 for the mass fraction of particulate matter that is below 10 μm.

[c] Cr^{+6} not measured concurrently with other metals.

[d] PM$_{10}$.

[e] One or more measurements below the detection limit.

TABLE 4-6. Average Emission Concentrations for HCl, PCDD, PCDF, and 2,3,7,8-TCDD Equivalents, Corrected to 12 percent CO_2 [Before Controls and (After Controls)][a]

Facility	% CO_2	HCl (g/dscm)	PCDD (ng/dscm)	PCDF (ng/dscm)	PCDD + PCDF (Total) (ng/dscm)	2,3,7,8-TCDD Equivalents (PCDD) (ng/dscm)	2,3,7,8-TCDD Equivalents (PCDF) (ng/dscm)	2,3,7,8-TCDD Equivalents (Total) (ng/dscm)
A[b]	6.2	3.12 (4.66E − 02 Sc.) (1.53 Bh.)	161.00 (19.83 Sc.) (2372.67 Bh.)	970.00 (94.47 Sc.) (2329.67 Bh.)	1,131.00 (114.30 Sc.) (4702.34 Bh.)	3.44 (2.35 Sc.) (23.37 Bh.)	28.16 (8.45 Sc.) (69.84 Bh.)	31.60 (10.80 Sc.) (93.21 Bh.)
B	6.3	1.74 (invalid data)	400.0 (246.67 Bh.)	1,086.99 (781.90 Bh.)	1,486.99 (1028.57 Bh.)	9.41 (6.84 Bh.)	42.99 (37.54 Bh.)	52.40 (44.38 Bh.)
C	5.6	0.88 (0.13 Sc.)	818.70 (283.97 Sc.)	1,705.50 (758.50 Sc.)	2,524.20 (1042.47 Sc.)	7.50 (3.87 Sc.)	51.46 (22.76 Sc.)	58.96 (26.63 Sc.)
D	4.4	3.39	1,009.09	2,030.45	3,039.55	9.44	71.54	80.97
E	7.0	1.89	106.48	255.99	362.47	1.26	10.27	11.53
F	6.3	1.80 (4.37E − 03 Sc.)	181.13 (9.46 Sc.)	499.25 (19.97 Sc.)	680.38 (29.43 Sc.)	1.70 (0.23 Sc.)	14.50 (1.44 Sc.)	16.20 (1.67 Sc.)
G	4.9	0.97	2,164.30	4,139.00	6,303.30	18.01	188.33	206.34
H	8.0	4.71	1,493.25	1,721.40	3,214.65	26.70	128.66	155.36

Sc. = scrubber; Bh. = baghouse.

[a] From individual source test reports.

[b] Data from before APCEs is taken at the scrubber inlet.

APC equipment are shown in parentheses. Actual emission rates downstream of APC equipment, if present, are summarized in Table 4–7.

Four of the incinerators tested were equipped with APC equipment.[3] The regional facility A was evaluated with *either* the wet scrubber *or* the dry lime scrubber with baghouse on line; the dry scrubber/baghouse was a backup system. Test engineers noted that the lime feed rate may not have been sufficiently high during testing, which could account for the relatively poor HCl control efficiency shown in Table 4–8. Other dry lime and spray-dried lime scrubbers with baghouses typically exhibit higher HCl removal efficiency (Teller et al. 1990; Lauber and Drum 1990).

Dioxin and furan analyses were drawn from different trains and at different times than the particle samples for metals. In the cases where the data were not drawn simultaneously, all data points were averaged for a given facility and the averages examined. This was thought to be a more representative procedure than attempting to relate individual data points that may have been obtained on different test days. A weighted least squares regression would have been appropriate for the averaged data, but in recognition of the fact that highly meaningful statistics were not being sought, and for convenience, only a simple regression was performed. Only linear regression models were attempted; in a few cases, a nonlinear model might have led to a significant correlation. Instances of linear regressions with an r^2 greater than about 0.5 were considered noteworthy. To reduce some of the equipment variability, all data reported were for samples drawn ahead of APC equipment, if present.

Regressions were performed on averages for all facilities and individual data points for the metals. Regression analyses for all metals against one another, TCDDEQV against HCl, PM_{10}, total metals, and individual metals were performed and are contained in Table 4–9 (Glasser 1990). For reasons to be discussed later, regressions were also performed with and without inclusion of facility A. Only those regressions and plots of the data deemed sufficiently unique to warrant mention are discussed in this chapter.

4.4 DISCUSSION OF RESULTS

The source test data on collection efficiencies of APCEs were in accord with the expectations discussed next.

---

[3] Afterburners were not considered to be APC equipment for the purpose of this study, since in some respects they acted as secondary combustion chambers.

TABLE 4-7. Emission Rates (in g/s) (After Controls, if They Exist)[a]

Facility	HCl	2,3,7,8-TCDD Equivalents	Particulate	Cd	As	Cr	Ni	Cr^{+6}	Pb	Hg
A [Sc.]	22.5E-3	5.59E-9	0.107	40.10E-6	1.37E-6	31.97E-6	13.6E-6[b]	3.52E-6	2189.E-6	129.E-6
A [Bh.]	641.3E-3	38.9E-9	0.0252	1.77E-6[b]	0.43E-6[b]	35.3E-6	34.26E-6	0.323E-6	42.6E-6	436.E-6
B	711.0E-3	23.43E-9	0.0042(PM_{10})	0.32E-6	0.002E-6[b]	0.13E-6	1.64E-6[b]	NM	48.6E-6[b]	NM
C	63.0E-3	31.27E-9	0.168	242.2E-6	4.97E-6	14.12E-6	11.02E-6[b]	0.717E-6[b]	1720.E-6	502000.E-6[c]
D	677.0E-3	39.78E-9	0.0025(PM_{10})	157.3E-6	5.92E-6	23.14E-6	11.76E-6	NM	6716.E-6	NM
E	316.0E-3	1.08E-9	0.015	3.11E-6	0.06E-6	13.6E-6[b]	1.21E-6[b]	0.003E-6[b]	38.6E-6	49.0E-6
F	2.7E-3	0.71E-9	0.0275(PM_{10})	60.52E-6	ND	11.E-6	6.04E-6[b]	10.47E-6	1992.E-6	NM
G	539.E-3	92.2E-9	0.101(PM_{10})	53.15E-6	0.35E-6[b]	9.04E-6[b]	10.23E-6[b]	NM	390.8E-6	NM
H	443.E-3	12.2E-9	0.0378	14.58E-6	0.13E-6	0.969E-6[b]	1.18E-6[b]	0.604E-6[b]	283.E-6	1370.E-6

ND = not detected; NM = not measured.
[a] Average of runs, from ARB source tests.
[b] One or more measurements are below detection limits.
[c] This data point is highly suspect.

TABLE 4–8. Air Pollution Control Device Summaries for Facilities with APCDs from ARB Source Tests[a]

Facility	Control Device	Manufacturer and Model[b]	HCl Control Efficiency 100[b] (In-Out)/In	2,3,7,8-TCDDEQV Control Efficiency 100[b] (In-Out)/In	Particulate Control Efficiency 100[b] (In-Out)/In	Cd Control Efficiency 100[b] (In-Out)/In	As Control Efficiency 100[b] (In-Out)/In	Cr Control Efficiency 100[b] (In-Out)/In	Cr^{+6} Control[c] Efficiency 100[b] (In-Out)/In	Pb Control Efficiency 100[b] (In-Out)/In	Hg Control Efficiency 100[b] (In-Out)/In
A	1. Main system venturi scrubber[d]	Emcotek Scrub #130	99.0	66.1	61.6	66.1	60.9	64.3	(−3.9)[e] 74.2 87.6	70.9	49.5
	2. Backup—lime injection scrubber with baghouse		44.9	(−203.0)[e] 31.9 90.6	87.6	98.5[f]	87.7	63.7	95.1	99.4	(−118)[e] (−68.6) 51.8
B	Fabric filter baghouse (total filter area of 1696 square feet)	Mikro-Pulsaire Baghouse	7.1	(−16.4)[e] 11.9	96.5	99.9[f]	99.9[f]	98.7	NM	99.8[f]	NM
C	Sodium hydroxide wet scrubber	Sly Wet Scrubber	81.7	48.7[e] (−3.4) 76.6	29.2[e] (−90) 81	14.5[e] (−434) 22.6	(−14.3)[e] (−474) 22.6	(−24.5)[e] (−45) 45.5	(−161) 67.7 40.9	1.0[e] (−92.3) 29.1	42.3[g]
F	Sodium hydroxide venturi scrubber and natural gas fired reheater	Ducon Wet Scrubber	99.7	87.7	(−61.3)[e] 6.94 48.5	(−62.5)[e] 48.6 36.9	ND	(−16.0)[e] 12.3 39	70.87[e] (−907) (−79.1)	(−185)[e] 4.5 30.2	NM

ND = not detected; NM = not measured.

[a] From individual source test reports, unless otherwise noted.
[b] Woodhouse, 1989.
[c] Cr^{+6} tests were not concurrent with other metals tests.
[d] Two systems do not operate concurrently.
[e] All test measurements are reported separately because of large spread.
[f] Exit measurements were below the detection limit.
[g] Two of three measurements, one was below the detection limits.

TABLE 4–9. Correlation Regression Data

Correlation	r^2 (complete data set)	r^2 (less facility A)	r^2 (complete data set, unaveraged)	r^2 (less facility A, unaveraged)
2,3,7,8-TCDDEQV—HCl	0.016	0.041	—	—
2,3,7,8-TCDDEQV—corr. PM$_{10}$	0.088	0.104	—	—
2,3,7,8-TCDDEQV—Total Metals	0.046	0.004	—	—
Total metals—corr. PM$_{10}$	0.086	0.368	—	—
2,3,7,8-TCDDEQV—As	0.049	0.037	—	—
2,3,7,8-TCDDEQV—Cd	0.036	0.036	—	—
2,3,7,8-TCDDEQV—Cr	0.102	0.093	—	—
2,3,7,8-TCDDEQV—Fe	0.000	0.524	—	—
2,3,7,8-TCDDEQV—Mn	0.079	0.014	—	—
2,3,7,8-TCDDEQV—Ni	0.022	0.039	—	—
2,3,7,8-TCDDEQV—Pb	0.065	0.002	—	—
As—Cd	0.738	0.746	0.382	0.241
As—Cr	0.044	0.125	0.060	0.263
As—Fe	0.022	0.003	0.004	0.071
As—Mn	0.071	0.306	0.101	0.079
As—Ni	0.001	0.041	0.035	0.272
As—Pb	0.075	0.127	0.083	0.011
Cd—Cr	0.015	0.208	0.013	0.223
Cd—Fe	0.013	0.066	0.016	0.022
Cd—Mn	0.053	0.708	0.013	0.389
Cd—Ni	0.007	0.019	0.019	0.103
Cd—Pb	0.111	0.536	0.205	0.479
Cr—Fe	0.828	0.003	0.368	0.092
Cr—Mn	0.934	0.134	0.165	0.608
Cr—Ni	0.884	0.489	0.823	0.425
Cr—Pb	0.834	0.096	0.561	0.084
Fe—Mn	0.855	0.04	0.222	0.079
Fe—Ni	0.799	0.131	0.330	0.000
Fe—Pb	0.879	0.361	0.414	0.103
Mn—Ni	0.815	0.105	0.585	0.261
Mn—Pb	0.897	0.438	0.427	0.281
Ni—Pb	0.786	0.162	0.595	0.049

1. Wet scrubbers were efficient in controlling HCl emissions.
2. Wet scrubbers tested, including one venturi scrubber, reduced but did not efficiently control fine particles, volatile metals (Hg) or dioxins and furans (which might have been present on fine particles and in the vapor phase).
3. Baghouses were efficient in controlling fine particulate matter and, consequently, the fine condensation aerosols produced by volatile metals.
4. Baghouses alone were not efficient in controlling the PCDDs and PCDFs and might have been an additional source of these compounds in the presence of adequate precursors and chlorine sources.
5. Baghouses alone were not efficient in controlling materials present in the vapor phase at the temperature of the baghouse.

The presence of readily measurable levels of PCDDs and PCDFs in *every* case indicates that combustion conditions, in the types of incinerators tested, were inadequate to prevent formation of PCDDs, PCDFs, or their precursors. The exact reasons for the penetration of PCDDs, PCDFs, or their precursors through the incinerators remain unclear from the test data, since CO and volatile organics were generally low, normally an indication of satisfactory combustion. Formation of soot particles in the primary chamber, perhaps during transients, and resulting in incomplete burnout because of insufficient secondary chamber residence time is a plausible hypothesis.

Some recent volatilization studies of polyvinylidene chloride (PVDC) show that precursor compounds to dioxins and furans such as chlorinated benzenes are major decomposition products of that polymer (Yasuhara and Morita 1988; Pasek and Chang 1991). The escape of PCDDs, PCDFs, or their precursors from the incinerator could lead to additional postcombustion formation and would be possible with such a scenario. Clearly, the two highest values for TCDDEQVs were associated with units G and H, both of which were reported to have very low primary chamber temperatures and were possibly accompanied by low temperature excursions in the secondary chamber. In the case of facility G, the presence of a heat exchanger in addition to poor combustion conditions might account for the highest levels of PCDDs and PCDFs measured. It is known that postcombustion formation of chlorinated arenes and PCDDs is thermodynamically favored at moderately low temperatures, i.e., in the range of 250 to 350°C (Chang, Mournighan, and Huffman 1991; Barton et al. 1990) and that such formation has been reported in laboratory experiments involving HCl or PCDD precursors using fly ash or metal-mediated oxidation-reduction reactions (Hagenmaier et al. 1987; Karasek and Dickson 1987; Eiceman et al. 1989; Hoffman et al. 1990; Bruce, Beach, and Gullett 1990).

The regression analyses of metals with TCDDEQVs did not yield any significant correlations when all data points were included. However, visual examination of the scatter diagrams indicated that a possible correlation was present between iron (Fe) and TCDDEQV as shown in Figure 4–2 (top). In fact with the single point from facility A included, there was no correlation; with that point removed, the correlation rose to 0.52, and removal of an additional point from facility G resulted in an increase of r^2 to 0.93, as shown in Figure 4–2 (bottom). Facility A was the regional waste facility that burned primarily "infectious waste" and had a much higher Fe content relative to PCDDs and PCDFs than the other units. As will be shown later, the high Fe probably resulted from large quantities of stainless steel sharps, e.g., needles and surgical implements fed to such a unit. Furthermore, a regional facility would be expected to have better trained operators, thereby minimizing the formation of PCDD and PCDF precursors, and since no heat recovery unit was present there was reduced likelihood of postcombustion formation. With the baghouse in place, TCDDEQV emissions for facility A rose. As noted above, facility G produced the largest quantities of TCDDEQVs, possibly because of its low primary temperature and the presence of a heat recovery unit. Therefore, it might have produced more TCDDEQV for a given amount of Fe in the ash because of a higher concentration of precursors. Recognizing that correlation does not imply causality and that there is a possibility of accepting a false-positive correlation, we suggest that this observation needs further investigation to clarify the relationship between Fe and PCDD and PCDF formation suggested by these data and by recent laboratory studies of MSW fly ash (Hoffman et al. 1990).

Since the metals analyses were determined from the same individual samples, facility averages and individual points could be related to one another. As seen in Table 4–9, several of the facility-averaged metals appeared to be correlated ($r^2 > 0.8$). However, the overall metals emissions from facility A for Fe, Mn (manganese), Ni (nickel), Cr (chromium), and Pb (lead) were high, so those points heavily weighted the regressions. The remaining facilities' data points were relatively closely clustered. For that reason, a high facility-averaged r^2 did not necessarily mean that each facility's metals concentrations were in the same proportion as for A. By plotting individual points, it became clear which metals' proportions were similar to one another. The strongest relation found was for Cr/Ni, $r^2 = 0.88$ (see Figure 4–3), followed by Fe/Pb and As/Cd (arsenic/cadmium) when all facilities were included (not shown). The correlation for Cd/Mn also improved to 0.7 when facility A was excluded.

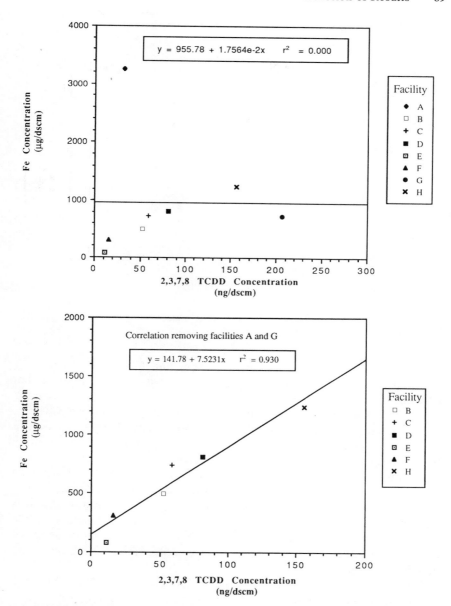

FIGURE 4–2. 2,3,7,8-TCDD equivalent concentration versus iron concentration, corrected to 12 percent CO_2, before APC equipment, if any.

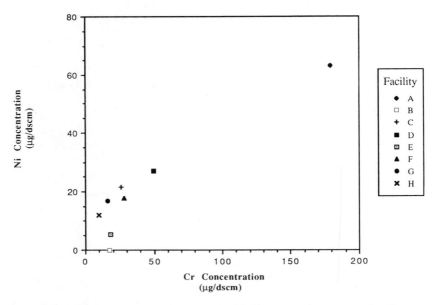

FIGURE 4–3. Chromium concentration versus nickel concentration, corrected to 12 percent CO_2, before APC equipment, if any.

We had not reported (Hickman, Chang, and Glasser 1989) that Cr was detected in some plastic items because it was not the focus of that investigation; however, we do so now. It is a common pigment (Hasselriis 1990). The earlier analyses also showed that Fe was present in such articles as well as being a component of stainless steel. The clear association of Fe and Cr strongly implicated stainless steel as a major metal-contributing waste component associated with the infectious waste stream. Recently, x-ray images of stainless steel particles in fly ash from a MWI have been reported, lending further credence to the observed correlation (Leger, Schlaegle, and Casuccio 1990). Thus the high concentration of stainless steel in the infectious waste stream, possibly augmented by metal chloride formation from chlorinated plastic waste containers and the presence of iron and chromium pigments, resulted in volatilization of these relatively high boiling point metals.

The reason for the association of Fe and Pb is less clear. We have previously reported (Hickman, Chang, and Glasser 1989) the presence of Pb either as a pigment or stabilizer in infectious waste containers (sharps and infectious waste bags). The relationship of Cd/Mn excluding facility A suggests predominant sources of Cd other than the infectious waste component. As reported earlier, Cd was observed in the outer casing of alkaline batteries used in disposable examination flashlights. Manganese (Mn) is also a constituent of alkaline batteries, the presence of which in hospital wastes can hardly be avoided if general wastes are not strictly segregated. It is possible that the weak correlation between Cd/Mn could be explained by such a source. A reason for the possible association of As/Cd is not known.

A final remark regarding Hg emissions is in order. The ARB source test program did not include Hg in every test, and with one exception, the Hg concentrations were not extraordinarily high. However, in one case, the reported emission rate was so large that it was thought that a typographical error had been made or that the sampling or analysis procedure was faulty. Discussion with the test engineer, however, confirmed that there was no typographical error, nor were any particular problems with that test noted. Mercury thermometers and batteries have largely been replaced, unlike crematories where Hg amalgams are known to be a source, so a source of Hg was somewhat of a mystery. Recently, we have learned that elemental Hg is still used as a weight in enteral feeding tubes (Popewiny 1991). Although recovery is practiced, there is little guarantee that recovery is complete, so it may be a significant source of Hg. Vigilance to prevent insidious introduction of metals into MWIs is needed.

4.5 SUMMARY

It is important that the composition of the waste stream and the conditions under which those materials are burned be better understood and controlled. Strategies capable of reducing dioxin and furan emissions, as well as those that can reduce metals emissions are necessary. Source reduction, recycling, and source separation are often touted as being favorable to end-of-pipe strategies, but these mitigation strategies have been neglected (Glasser et al. 1991). If elements of the MW stream can be identified as primary sources of specific toxic emissions, an alternative to meeting best available control technology-driven, risk-based standards would be elimination of those elements from the combustible component.

Metals stand out as the primary example of materials that not only persist but, because of their nucleation and condensation properties, tend to form fine particulate matter. The health risk posed by these metals becoming airborne can be exacerbated by efficient combustion. The most effective mechanism to reduce such toxic combustion products would be to eliminate products containing those constituents from the MW stream. The simplest mechanism to eliminate products from the waste stream would be to return to a reliance upon reusable and recyclable products and autoclavable surgical implements and sharps. In cases where these measures are not feasible, the elimination of these toxic elements from the waste stream, depending upon whether the source is diffuse or concentrated, could occur through product substitution or product reformulation. If no less toxic alternative exists, these materials could at least be segregated from the fraction of the waste stream that is to be incinerated.

4.6 RECOMMENDATIONS

Several steps to reduce metals emissions from MWIs are suggested by the analyses. The material composition of MW is diverse and may depend upon daily hospital activities. A more useful system of classification, from a public health standpoint, would emphasize the chlorine and metals content of the items in the waste stream. This scheme would necessitate labeling (similar to that in the food industry) and would require large-scale cooperation from the manufacturers of medical supplies. It has been argued that a detailed sampling and chemical analysis of the varied constituents that make up the MW stream may be necessary to determine which elements pose the greatest potential risk upon incineration. But this roundabout manner of determining product composition would be necessary only in the absence of vendor-supplied data.

Knowledge of the offending components can lead to reduction of metals inputs to the incinerator by alternative disposal. For example, metallic sharps or electronic devices such as pacemakers could be sterilized, followed by grinding or solidification (some sharps containers are now designed that way), and landfill disposal. Taking steps of that kind should reduce Cr^{+6} emissions to negligible levels if chromium-containing pigments are simultaneously restricted. Strict segregation programs are also possible (for example, eliminating disposable flashlights in favor of alternatives, keeping general wastes out of the incinerator, and requiring manufacturers to reformulate products, providing certification that products destined for the infectious waste stream do not contain Pb, Cd, or Cr). Reduced dependence upon chlorinated plastics would simultaneously reduce HCl emissions, by developing substitutes for PVC IV tubing and blood bags and substituting clear polyethylene or cellulosic waste bags (Bulley 1990; Green et al. 1990a). Increased operator training or certification would improve waste feeding practices and equipment maintenance to enhance combustion, limiting the possibility of PCDD, PCDF, or precursor formation during combustion (Pinder 1990).

Our research suggests that many useful technical and policy suggestions can be gleaned from an expanded perspective. If the goal of hospitals is to provide competent and effective health care, it makes sense to ask why metallic elements are entering the MWIs and whether they need to be there, before emphasizing improved APC equipment. While we would like to stress the utility of this broad-based approach, we think that future research should also be directed toward understanding the apparent metals (Fe) correlation with PCDD and PCDF formation, and whether upstream HCl control can effectively reduce downstream PCDDs and PCDFs, as has been suggested by Green et al. (1991).

ACKNOWLEDGMENTS

The authors are indebted to the staff of the Air Resources Board: Bob Adrian, Barbara Fry, Kitty Howard, Roland Hwang, Al Jenkins, George Lew, Gloria Lindner, James McCormack, Peter Ouchida, Pat Randall, and Luis Woodhouse. Without their cooperation and the use of their source test and hosptial survey data, this study would not have been possible. Partial support for this project was provided by the University of California, Davis, Public Service Research and Dissemination Program, and the National Institute of Environmental Health Sciences grant ES-04699-0151 (Superfund Basic Research Program).

REFERENCES

Anonymous. 1990. *Waste Age* (November): 101.
Barton, R. G., G. R. Hassel, W. S. Lanier, and W. R. Seeker. State-of-the-art assessment of medical waste thermal treatment. Report for the U.S. EPA Contract #68-03-3365 and California Air Resources Board Contract #A832-155, Energy and Environmental Research Corporation, Irvine, CA, 1990.
Barton, R. G., and W. R. Seeker. "Behavior of metals in medical waste incinerators." Paper No. 90-38.5 presented at the 83rd Annual Meeting and Exhibition of the Air and Waste Management Association, Pittsburgh, June 24–29, 1990.
Beychok, M. R. 1987. A data base of dioxin and furan emissions from municipal refuse incinerators. *Atmospheric Environ.* 21: 29–30.
Bruce, K. R., L. O. Beach, and B. K. Gullett. "The role of gas-phase Cl_2 in the formation of PCDD/PCDF in municipal and hazardous waste combustion." Paper No. 5.7.1 presented in the Proceedings of the Incineration Conference 1990, Hanalei Hotel, San Diego, CA, May 14–18, 1990.
Brunner, C. R. and C. H. Brown. 1988. Hospital waste disposal by incineration. *J. Air Pollut. Control Assoc.* 38: 1297–1309.
Bulley, M. M. "Medical waste incineration in Australasia." Paper No. 90-27.5 presented at the 83rd Annual Meeting and Exhibition of the Air and Waste Management Association, Pittsburgh, June 24–29, 1990.
Carroll, G. and L. R. Waterland. "Parametric evaluation of metal partitioning at the U.S. EPA incineration research facility." Presented at the Risk Reduction Research Laboratory Symposium, Cincinnati, OH, May 1990.

94   Avoidance of Metals in Biomedical Waste Incinerators

Chang, D. P. Y., R. E. Mournighan, and G. L. Huffman. 1991. An equilibrium analysis of some chlorinated hydrocarbons in stoichiometric to fuel-rich post-flame combustion environments. *J. Air Waste Manage. Assoc.* 41: 947–955.

Doyle, B. W., D. A. Drum, and J. D. Lauber. 1985. The smoldering question of hospital wastes. *Pollut. Eng.* 16: 35–39.

Drum, D. A. 1990. Butler County Community College, Butler, PA, personal communication.

Eiceman, G. A., et al. "The chlorination of aromatic compounds adsorbed on fly ash surfaces by hydrogen chloride: Experimental evidence for environmental contamination by waste incineration." Preprint, Division of Environmental Chemistry, American Chemical Society, Dallas, April 9–14, 1989.

Glasser, H. "Hazardous emissions from hospital waste incinerators: Issues, status, and mitigation measures." Master of Science Thesis, Department of Mechanical Engineering, University of California, Davis, May 1990.

Glasser, H., D. P. Y. Chang, and D. C. Hickman. 1991. An analysis of biomedical waste incineration. *J. Air Waste Management Assoc.* 41: 1180–88.

Green, A. E. S., et al. "Toxic products from co-combustion of institutional waste." Paper No. 90-38.4 presented at the 83rd Annual Meeting and Exhibition of the Air and Waste Management Association, Pittsburgh, June 24–29, 1990.

Green, A. E. S., J. C. Wagner, and K. J. Lin. 1991. Phenomenological models of chlorinated hydrocarbons. *Chemosphere* 22: 121–25.

Greenburg, R. R., W. H. Zoller, and G. E. Gordon. 1978. Composition and size distribution of particles released in refuse incineration. *Environ. Sci. Technol.* 12: 566–73.

Hagenmaier, H. et al. 1987. Catalytic effects of fly ash from waste incineration facilities on the formation and decomposition of polychlorinated dibenzo-*p*-dioxins and polychlorinated dibenzofurans. *Environ. Sci. Technol.* 21: 1080–4.

Hasselriis, F. "Relationship between waste composition and environmental impact." Paper No. 90-38.2 presented as the 83rd Annual Meeting and Exhibition of the Air and Waste Management Association, Pittsburgh, June 24–29, 1990.

Hickman, D. C., D. P. Y. Chang, and H. Glasser. "Cadmium and lead in biomedical waste incinerators." Paper No. 89-110.6 presented at the 82nd Annual Meeting and Exhibition of the Air and Waste Management Association, Anaheim, CA, June 25–30, 1989.

Hoffman, R. V., et al. 1990. Mechanism of chlorination of aromatic compounds adsorbed on the surface of fly ash from municipal incinerators. *Environ. Sci. Technol.* 24: 1635–41.

Huffman, G. L. "Environmental controls for waste-to-energy plants." Presented at the USDOE/TVA/Argonne/Kentucky Energy and Environmental Protection Cabinets Conference on *Energy from Solid Waste: An Option for Local Governments,* Louisville, KY, May 13–15, 1987.

Jenkins, A. Evaluation test on a hospital refuse incinerator at Cedars Sinai Medical Center, Los Angeles, CA. Air Resources Board Test Report No. SS-87-11, 1987a.

Jenkins, A. "Evaluation test on a hospital refuse incinerator at Saint Agnes Medical Center, Fresno, CA." Air Resources Board Test Report No. SS-87-01, 1987b.

Jenkins, A. "Evaluation test on a refuse incinerator at Stanford University Environmental Safety Facility, Stanford, CA." Air Resources Board Test Report No. ML-88-025, 1988a.

Jenkins, A. "Evaluation retest on a hospital refuse incinerator at Sutter General Hospital, Sacramento, CA." Air Resources Board Test Report No. C-87-090, 1988b.

Jenkins, A. "Preliminary test data of American Environmental Management Corporation Medical Waste Incinerator, Sacramento, CA." Air Resources Board Draft Test Report, November 17, 1989.

Johnson, N. 1990. Personal communication.

Karasek, F. W., and L. C. Dickson. 1987. Model studies of polychlorinated dibenzo-*p*-dioxin formation during municipal refuse incineration. *Science* 237: 754–6.

Lauber, J. D., and D. A. Drum. "Best controlled technologies for regional biomedical waste incineration." Paper No. 90-27.2 presented at the 83rd Annual Meeting and Exhibition of the Air and Waste Management Association, Pittsburgh, June 24–29, 1990.

Lee, C. C., G. L. Huffman, and T. L. Shearer. A review of biomedical waste disposal: Incineration." Fact Sheet, U.S. EPA, Office of Research and Development, Cincinnati, OH, 1988.

Lee, C. C. and G. L Huffman. 1991. Medical waste management. *Environ. Sci. Technol.* 25(3): 360.

Leger, M., J. R. Schlaegle, and G. S. Casuccio. "The use of microimaging to characterize waste incinerator by-products." Paper No. 90-38.3 presented at the 83rd Annual Meeting and Exhibition of the Air and Waste Management Association, Pittsburgh, June 25–29, 1990.

Marrack, D. 1988. Hospital red bag waste: An assessment and management recommendations. *J. Air Pollut. Control Assoc.* 38: 1309–11.

McCormack, J. E. "Evaluation test on a small hospital refuse incinerator, Saint Bernadines Hospital, San Bernardino, CA." Draft Report, Air Resources Board Test Report No. C-87-092, 1988a.

McCormack, J. E. "Evaluation test on a small hospital refuse incinerator, LA County—USC Medical Center, Los Angeles, CA." Preliminary Draft. Air Resources Board Test Report No. C-87-122, 1988b.

McCormack, J. E. "Preliminary data from evaluation test on a hospital refuse incinerator at Kaiser-Permanente Hospital, San Diego, CA. "ARB Test Report No. C-88-013, 1988c.

Morrison, R. "Hospital waste combustion study data gathering phase." Prepared by Radian Corporation, Research Triangle Park, NC, under U.S. EPA Contract No. 68-02-4330, 1987.

Oppelt, T. E. 1987. Incineration of hazardous waste: A critical review. *J. Air Pollut. Control Assoc.* 37: 558–86.

Pasek, R. and D. P. Y. Chang. "Potential benefits of polyvinyl chloride and polyvinylidene chloride reductions on incinerator emissions." Paper No. 91-33.1 presented at the 84th Annual Meeting and Exposition of the Air and Waste Management Association, Vancouver, June 23–28, 1991.

Pinder, S. N. "Observed effects on efficiency of operating hospital incinerators created by in-feeding a consistent, standard-densified, waste fuel." Paper 14.3.1 presented at the Proceedings of the Incineration Conference 1990, Hanalei Hotel, San Diego, CA, May 14–18, 1990.

Popewiny, S. Heavy metals in incinerators. Chem Alert Article, Center for Healthcare Enviromental Management, Plymouth Meeting, PA, 1991.

Ramsey, C. 1990. California Air Resouces Board, personal communication.

Teller, A. J., J. Y. Hsieh, P. Koch, and A. Astrande. 1990. Emission Control Hospital Waste Incineration. Preprint, personal communication.

U.S. EPA. "Hospital waste combustion study—data gathering phase—final report." Contract No. 68-02-4330, U.S. EPA Office of Air Quality Planning and Standards, Research Triangle Park, NC, 1988.

Woodhouse, L. 1989. California Air Resources Board, biomedical waste incinerator/raw survey responses, personal communication.

Yasuhara, A. and M. Morita. 1988. Formation of chlorinated aromatic hydrocarbons by thermal decomposition of vinylidene chloride polymer. *Environ. Sci. Technol.* 22(6): 646.

5

Relationship between Input and Output

Floyd Hasselriis

5.1 INTRODUCTION

Incineration of medical wastes has been the traditional method for disposal of infectious wastes from hospitals and other health care facilities. In the past, most hospitals were provided with furnaces capable of destroying at least pathological and infectious waste. Many also have sufficient capacity to burn a large part or all of the general waste produced at the hospital. The cost of off-site disposal and legal responsibility for proper disposal encourage hospitals to continue to use incineration.

Concern about the emissions from old and often primitive incinerators has resulted in more stringent regulation. States that once required only periodic inspection for opacity now require annual testing and have standards for particulate and HCl emissions from medical waste incinerators (MWIs). The U.S. EPA is currently developing national standards for MWIs.

Concern about the high levels of HCl emissions from combustion of hospital waste has led to efforts to reduce the quantity of chlorinated materials in hospital wastes. Likewise, concern about the presence of potentially toxic heavy metals such as lead, cadmium, and mercury in combustion emissions has increased interest in reducing the amount of these metals in the materials purchased by hospitals.

Various steps can be taken to prevent the discharge of unacceptable quantities of potentially harmful substances to the environment:

- Don't make it.
- Don't buy it.
- Send it elsewhere.

- Remove objectionable materials.
- Remove valuable (recyclable) materials.
- Burn properly.
- Limit emissions.
- Disperse stack gases.

The objective is to make certain that

- Acceptable ambient concentrations of critical substances do not approach (not to mention exceed) maximum safety levels from an environmental health point of view.
- General levels of pollutants in the atmosphere are not increased significantly by the emissions from incinerators.

To determine whether the emissions from a specific incinerator application are acceptable, it is necessary to know the *quantities* of the substances in the emissions and the *concentrations* in the gases leaving the incinerator and as they arrive on the ground or at a higher-level receptor. We need to know how closely the calculated concentrations approach acceptable health standards in order to know if the emissions are significant. We should not waste time and money on insignificant emissions.

On the basis of the extensive knowledge obtained from testing, we know we have to focus on the local and regional effects of acid gases, carbon monoxide, and organic emissions, as well as acceptable local and regional effects of heavy metals. Concern about particulate matter has shifted to the metals contents rather than the absolute amount of the particulate.

To "clean" the waste, we need to find the actual sources of the unwanted emissions and understand to what extent we can control their path through the process.

The gaseous and particulate emissions as well as collected fly ash and bottom ash residues from combustion of both municipal and medical solid wastes are related to the chemical and mineral composition of the waste, the combustion process, and the devices used to control and collect the emissions.

Ideally, we would chemically analyze each item in the waste and determine what happens to these chemicals and minerals, how they interact in combustion and in the emission control system, and in what final chemical form they leave the stack or end up in the landfill where the ash goes.

Actually, we know a great deal about almost all of these processes, as well as how to design and control combustion and emission control systems. We know the general composition of the waste, and we have the results of extensive stack testing. However, we need to put the whole picture together in order to understand what in the input produces this output.

In this chapter we will investigate the sources of acids, metals, and organics in the waste, what happens during combustion, what are the uncontrolled and controlled emissions, what factors affect the dispersion of these pollutants in the atmosphere before reaching ground level, and how the ground level and regional effects of these emissions are evaluated and compared with health standards.

There are many ways to approach environmentally sound clean combustion. Reduction, recycling, and good technology are basic. Selection of the appropriate technology includes not only incinerator design and emissions controls, but also a decision as to how high the stack should be to obtain effective dispersion at a specific site.

Other decisions must be made: Should only pathological waste be burned, since it is benign? Should only infectious waste be burned, or should general waste be added to keep the incinerator going? Should the infectious waste be sterilized before shipping it off-site to an incinerator?

Removing potentially polluting materials from the medical waste stream can reduce emissions from an incinerator. This procedure involves some effort and cost, which may not be justified if the incinerator already has emission controls that can remove almost all harmful components. On the other hand, if a clean waste can be obtained by removing harmful substances and toxics from the waste before it is burned, it may be possible to reduce the cost of the emission control system, or it may not be needed. If the emission control system exists but has limited efficiency, cleaning the waste may make it unnecessary to upgrade the emission control system.

In summary, estimated effects on the environment should be the basis for determining which, if any, pollutants require interception or control. This chapter will trace the inputs of major pollutants of concern from the waste to the output, relating the emissions to estimated environmental and health effects.

5.2 REGULATORY LIMITS ON EMISSIONS

National standards limit ground level concentrations of criteria pollutants as shown in Table 5–1. Hydrochloric acid is not included, nor are dioxins.

The states are developing and revising regulations applying to medical or biological waste incinerators. Different standards apply to existing MWIs than to new installations. Upgrading of existing MWIs may be required within various time periods.

A recent survey of current and/or anticipated regulations shows that most states have had, and retain, staging of requirements in accordance

TABLE 5–1. National Air Quality Standards (NAQS)

Contaminant	Averaging Period	National ($\mu g/m^3$) (Primary)
Carbon monoxide	8 h	10,000 (9 ppm)
	1 h	40,000 (35 ppm)
Sulfur dioxide	Annual	80 (0.03 ppm)
	24 h	365 (0.14 ppm)
Nitrogen dioxide	Annual	100 (0.05 ppm)
Total suspended particulates	Annual	75
	24 h	260
Fine particulates	Annual	50
	24 h	150
Lead	3 months	1.5

Source: U.S. EPA. *National primary and secondary Ambient Air Quality Standards*. 40 CFR 50.

with specific ranges in charging rate for emissions of total suspended particulate matter (TSP) and HCl. Table 5–2 provides a rough summary of regulations throughout the United States.

About 15 percent of the states have no regulations for HCl. About 1 percent regulate on a case-by-case basis, several using health risk basis; 40 percent limit emissions to 4 lb/h, or 90 percent control, or 50 ppm by volume; and about 1 percent require 95 to 99 percent control, or 30 ppm

TABLE 5–2. Ranges of Regulations for TSP and HCl in the United States

Pollutant	Charging Rate (lb/h)		
	(100–500)	(200–1,000)	(1,000–4,000)
Particulate (gr/dscf)	0.20	0.04	0.020
	0.08	0.02	0.010
Hydrochloric acid (ppmv)	4 lb/h	90%	50 ppm by volume
		99%	30 ppm by volume

Source: Hasselriis et al. 1991.

by volume. Generally, states with higher population densities have more stringent requirements. The limit of 4 lb/h is the emission of a 1,000 lb/h hazardous waste incinerator burning a high-chlorine waste such as polyvinyl chloride (PVC).

The U.S. EPA originally set numerical emission limits for hazardous waste incinerators of 0.08 grains per dry standard cubic foot (gr/dscf) and 4 lb/h or 99 percent control of HCl. Recently risk-based regulations have been added: Emissions of toxic metals are now limited on the basis of ground level health risk concentrations. Acceptable ground level concentrations (GLCs) are related back to stack concentrations by dispersion modeling in order to define acceptable stack emissions. A table of allowable stack concentrations is used for the first screen, based on waste composition and a conservative dispersion model. If the applicant expects to exceed these numbers, modeling based on the specific site and technology is required in order to obtain a permit to construct, and trial burn tests must be performed. Health risk–based regulations for MWIs have been developed by several states following the U.S. EPA approach for hazardous waste incinerators, focusing on specific pollutants, including toxic heavy metals.

5.3 ACCEPTABLE GROUND LEVEL AIR QUALITY STANDARDS

The environmental health impact is the logical starting point for an analysis of pollutants from incineration of wastes. The federal government and the states have developed requirements based on health risk in the workplace: the acceptable Ambient Air Quality Standards (AAQS), which are the primary basis for evaluating emissions from small generators.

When large generators that impact the regional environment are used, the fraction of criteria pollutants the facility could contribute to the background regional environment must also be considered. Absolute emissions levels, not necessarily based on health risk, are also imposed for many pollutants, and best available control technology (BACT) may be required in noncomplying regions, typically large urban areas.

Working from Air Quality Standards Back to the Waste

Ground level concentrations (GLC) that would result form emissions from an incineration facility are calculated by computer modeling. These calculations take into account the height of the stack, the gas temperature and velocity, and the configuration of the incinerator building and other structures on- and off-site. The GLCs of specific pollutants are compared with the AAQS to determine whether any of the standards are exceeded. The concentrations permitted to be emitted from the stack of an incinerator can be determined by working back the acceptable GLC to the stack by the same modeling procedure or by an arbitrary screening model. This procedure is often used to calculate the maximum allowable stack concentrations (MASC) permitted for a facility.

Based on the MASC, it is possible to determine the quantities of specific pollutants that might be allowed in the waste to be burned. It may be assumed, based on recent U.S. EPA tests, that all the chlorine, and about 75 percent of the cadmium and 50 percent of the lead in the waste will be found in the combustion gases (Durkee and Eddinger 1991). If knowledge of the composition of the waste indicates that certain potential pollutants in the waste would exceed the MASC, two avenues may be explored: removing these pollutants from the waste and removing them from the gases after they exit the incinerator.

The procedure described above, "going from Z to A," is not generally followed. On the contrary, absolute emission limits are often imposed in place of a MASC based on health risk. Best available control technology may be required in urban or other nonattainment areas on the basis that any emission is too much. The benefits of removing toxics from the waste stream may not be taken into account. Use of BACT, regardless of health and environmental effects, may be unnecessarily costly, raising the cost of disposal and the cost of health care.

5.4 CHEMICAL COMPOSITION OF WASTE

The chemical composition of the waste to be burned can be estimated by grinding up representative samples and analyzing them. This method determines the total composition of the wastes, including acid-forming sulfur, chlorine, and nitrogen, heavy metals, trace organics in the wastes, such as chlorophenol, chlorobenzene, and even dioxins and furans. However, this method does not reveal which components of the waste make the most significant contributions to emissions (NITEP 1985).

A more direct method for estimating emissions from typical waste types is to perform detailed testing of actual *uncontrolled* stack emissions from a facility having appropriate combustion technology. The California Air Resources Board (CARB) test program (described in Chapter 4) brought to light the performance of MWIs of various designs and vintages, with and without emission controls. These data include uncontrolled as well as controlled emissions and both poor and fairly good performance of emission controls (Fry et al. 1990). Recent U.S. EPA test data greatly expand the available data as well as the effect of operating conditions on emissions (Durkee and Eddinger 1991).

The amount of HCl and SO_2 measured in the stack reveals the amount of chlorine and sulfur in the waste. Also, the quantities of heavy metals such as lead, cadmium, and mercury released from burning the combustible materials can be measured by analysis of the particulate matter emitted and converted to *emission factors,* expressed as pounds per ton of waste burned. The fraction of the metals remaining in the ash residues remains unknown, but is of no interest from an emissions point of view. Neither of these methods pinpoints which components of the waste generated the toxics so that they can be targeted for action. For this reason, many generalities are rampant, such as blaming "the plastics" without identifying which ones. Analysis of the specific components of waste is necessary in order to identify the contributors of unwanted and toxic substances to the emissions.

The approximate total composition of typical medical waste and the products of combustion from a typical starved-air incinerator are shown in Table 5–3. Two-thirds of the products consist of nitrogen added with the oxygen required for combustion and excess air needed to control the temperature of the furnace.

The ash and inerts are assumed to be the bottom ash. The metals in the

TABLE 5–3. Chemical Composition of Waste and Combustion Products

Composition	%	lb/t	Products of Combustion	lb/t
Carbon	33.3	666	Carbon dioxide	1,693
Hydrogen	5.9	118		
Oxygen	11.5	230	Oxygen	512
Chlorine	1.0	20	HCl	20
Nitrogen	3.0	6	Nitrogen	5,820
Sulfur	2.0	4	Sulfur dioxide	8
Water	23.5	470	H_2O	1,127
Inerts	24.3	486	Flyash	3
			Residues	486
Lead			Lead	0.057
Iron			Iron	0.007
Cadmium			Cadmium	0.0066
Manganese			Manganese	0.0004
Chromium			Chromium	0.00036
Arsenic			Arsenic	0.00028
Total:		2,000		9,099.07164

Source: Hasselriis 1990.

combustibles in this case have been measured in the stack gases and are assumed to be present in these quantities in the combustible materials. This typical medical waste analysis shows that 1 percent chlorine would produce 20 lb/t of HCl in the incinerator emissions. This converts to about 1,000 ppm by dry volume. The ash consists mostly of inert materials, but 0.07164 lb/t of metals attracts our attention. These would be found in the fly ash.

5.5 COMPONENT COMPOSITION OF MUNICIPAL AND MEDICAL WASTE

The sources of pollutants in the waste can be identified by analyzing the individual components of the waste. The weight contribution of each component is used to determine the contributions from all components.

The component composition of typical municipal solid waste (MSW) and medical waste (MW) is shown in Table 5–4, showing substantial differences in some of their components. They contain similar amounts of paper. The yard waste in MSW has a counterpart in fluids in MW. The main difference is that MW contains two to three more times the amount of plastics than municipal waste. Medical wastes contain a large amount of disposable items made of plastics, whereas MSW contains a variety of plastic containers and consumer items, including electronic components and batteries.

Chemical Analysis of the Waste Components

The individual components of the waste are analyzed for chlorine, sulfur, and nitrogen, which determine the quantity of acids produced by combus-

TABLE 5–4. Composition of Red-Bag, Clear-Bag, and MSW

Composition	Red-Bag[a] (%)	Clear-Bag[a] (%)	MSW[b] (%)
Paper	31.0	36.0	35.6
Cardboard	0.0	3.0	10.4
Plastics	29.0	20.0	9.5
Rubber	12.0	1.4	1.4
Textiles	5.0	2.1	2.5
Food	1.0	11.7	10.1
Yard waste	0.0	2.0	15.0
Glass	3.2	4.8	6.4
Metals	1.1	7.2	5.1
Fluids	17.7	9.9	0.0
Misc. organics	0.0	1.9	5.4
Total	100.0	100.0	100.0

[a] From Brown 1989.
[b] From Hasselriis 1984.

tion. The ash is analyzed to quantify the metals, especially the heavy metals, which are of concern when emitted after combustion or when disposed of in ash residues.

Metals Sources in Components of Municipal and Medical Waste

The main component of MSW and MW is paper. Table 5–5 shows that pure and recycled paper, including magazines and junk mail, contain measurable levels of lead, cadmium, and mercury. Lead content ranges up to 2.6 ppm and cadmium up to about 0.4 ppm by weight, and significant amounts of mercury are found.

The metals content of municipal wastes and composts, shown in Ta-

TABLE 5–5. Lead, Cadmium, and Mercury in Wastes

Component	Lead (ppm by wt)	Cadmium (ppm by wt)	Mercury (ppm by wt)
Pine wood	0.20	0.23	
Groundwood			
unbleached	1.28	0.13	
bleached	1.79	0.21	
Paper			
upper range	2.56	0.41	0.13
lower range	0.32	0.10	0.02
Newspaper			
unprinted	0.32	0.13	
comics	0.37	0.07	
classified section	0.48		
slick advertising	0.72	0.10	
slick advertising	0.79	0.30	
bond paper	0.50	0.05	
Junk mail, glossy	0.6	0.04	
Junk mail, glossy	0.92	0.03	
Magazine, glossy	0.90	0.05	
Slick advertising	1.30		
Commercial recycled	0.76	0.4	0.11
Commercial recycled	2.56		
Residential recycled	2.2		
Residential recycled	2.5	0.1	0.05

Sources: Kerstetter and Lyons, 1989; Hasselriis 1990.

ble 5–6, are much higher than those of the paper fraction, indicating that other sources are more significant. The lead content of compost made from trees and shrubs tested in Europe was found to be 10 times as high as in wood and paper and similar to MSW, possibly due to the continued use of lead in gasoline. The compost made from MSW contained 560 ppm by weight of lead, 200 times that of paper, and 3.8 ppm of cadmium. Refuse-derived fuel was found to contain an average of 160 ppm of lead, and to have 1.6 ppm of cadmium and almost 1 ppm of mercury.

Recent tests of municipal waste components, shown in Table 5–7, show surprisingly high levels of lead, chromium, and nickel, especially in certain plastics found in household waste. The *uncontrolled* stack emission of incinerators burning municipal and medical wastes are compared in this table.

The stack emissions from MWIs are remarkably similar and are substantially lower than those resulting from burning MSW. Note that the North Little Rock and Prince Edward Island plants, which have modular, starved-air technology similar to that used by most medical waste combustors, show substantially lower emissions than the Commerce facility, which has excess-air, waterwall technology. The lead and cadmium emissions from medical waste combustion were consistently lower than those from municipal waste.

TABLE 5–6. Lead, Cadmium and Mercury in Wastes

Component	Lead (ppm)	Cadmium (ppm)	Mercury (ppm)
Municipal waste			
Prince Edward Island	60	3.0	3.0
Portland, OR, residential	24	1.0	
Portland, OR, commercial	50	1.2	
Des Moines, IA	33	0.5	0.9
Swedish data	37	1.1	1.5
Average	44	1.5	1.8
Compost, tree and shrub	23	<1	<1
Compost, Garden	57	0.9	<1
Compost, paper	53	2.8	0.1
MSW compost (13 plants)	560	3.8	2.9
RDF, 15-plant average	160	1.6	0.9
U.S. Bureau of Mines MSW	330	9.0	

Sources: Hasselriis 1984; Bergvall and Hult 1986; U.S. EPA 1987.

TABLE 5–7. Metals Content of Municipal Waste Components (ppm)

Components	Lead	Cadmium	Chromium	Nickel
Office paper	75	1.7	270	800
Mixed paper	50	1.3	160	500
Magazines	80	1.6	300	540
Diapers	60	1.7	90	370
PET bottles	400	40	200	480
HDPE bottles	350	1	120	300
EPS	50	1	200	1,700
Other plastics	160	13	700	1,000
Plastic containers	100	8	800	1,500
Calculated from emissions[a]				
Municipal waste incinerators[b]				
Commerce, CA	154	8.2		
North Little Rock	67	2.0		
Prince Edward Island	60	3.0		
Medical waste incinerators[c]				
Sutter General	11.9	0.9		
Stanford	23.1	0.8		
Kaiser-Permanente	21.4	1.4		
St. Agnes	21.7	1.6		
Cedars Sinai	21.7	2.2		

PET = polyethylene terephthale; HDPE = high density polyethylene; EPS = expanded polystyrene
[a] From Hasselriis 1990.
[b] From U.S. EPA 1987.
[c] From Morrison 1987.

The levels of lead and cadmium in MSW and MW emissions reflect the higher metals contents of the paper and plastics components, due to large amounts of fillers and colorants. Table 5–8 lists the metals used in printing inks and as colorants and stabilizers (fillers) in plastics.

Chlorine Sources in Municipal and Medical Waste

Studies of municipal waste have shown that about half the chlorine content of municipal waste comes from chlorinated plastics. The remainder is found mainly in the paper and food wastes. Medical waste has a higher plastics content and contains much less food.

The main contributor of chlorine in both MSW and MW is PVC. Since it contains about 45 percent chlorine, it can contribute a large fraction of the HCl emissions. Hospital waste may also include chlorinated solvents and disinfectants.

The proportion of various types of plastics in medical waste described

TABLE 5–8. Metals Used in Papers and Plastics

Metals Used in Printing Inks[a]
  Chrome yellow: chromium combined with lead
  Chrome green: chrome yellow and iron blue
  Molybdate orange: molybdenum and lead
  Cadmium-mercury: bright to deep red shades

Metals Used as Stabilizers in Plastics[b]
  Barium/cadmium
  Barium/zinc
  Barium, lead
  Barium
  Cadmium
  Calcium
  Calcium/zinc
  Lead
  Magnesium
  Phosphates
  Strontium
  Tin, barium
  Tin, calcium

Metals Used as Colorants[b]
  White: Titanium dioxide, Zinc sulfide, Zinc oxide
  Red, red–orange: Cadmium, iron, chrome, cobalt
  Orange, yellow, buff: Cadmium, chrome, molybdenum, nickel, iron
  Blue-green: Cobalt, chrome
  Pearl: Lead carbonate

[a] From the National Association of Printing Ink Manufacturers, Inc.
[b] From the *Plastics Encyclopedia,* McGraw-Hill, New York, 1986.

in Table 5–4 is shown in Table 5–9 (Brown 1989). Note that since the red-bag (infectious) waste contained 29 percent plastics, and 15 percent of that is PVC containing 45 percent chlorine, this waste contains 2 percent chlorine. The emission factor for the red-bag waste is thus 40 pounds of chlorine per ton of waste. Without acid gas controls, this would produce emissions of about 1,300 ppm by volume and would require 96 percent control to reduce the emissions to the 50 ppm required for large incinerators. The clear-bag (noninfectious) hospital waste with 18 percent PVC at 20 percent of the plastics would produce about 30 lb/t, or about 1,000 ppm by volume of HCl.

Municipal waste with 9.5 percent plastic, of which about 10 percent might be PVC, would contain less than 9 pounds of plastic-derived chlorine per ton, contributing less than 300 ppm by volume of HCl. Since general hospital waste is similar to municipal waste, it may also be expected to have a lower chlorine content. Burning general waste with infectious waste would thus tend to reduce the average HCl emissions.

TABLE 5–9. Plastic Types in Hospital Waste

Component	Red-Bag (%)	Clear-Bag (%)
Polyethylene	45	36
Polypropylene	15	14
PVC	15	18
Polycarbonate	11	14
Polystyrene	4	10
Mixed	11	9

Source: Brown 1989.

5.6 EFFECTS OF THE COMBUSTION PROCESS

The first rule of combustion is that what goes in must come out. It is not easy to determine in what chemical form it will appear, however. All of the acid-forming chlorine, sulfur, and nitrogen are expected to be emitted in the combustion gases. The fate of other elements is not as simple to determine.

Environment Canada has carried out extensive tests on several MSW combustion systems, one a modular (starved-air) system and one a mass-burn waterwall system. The data from these tests provide a good input-output mass-balance of the organic and inorganic components of MSW. These tests provide detailed information on the fate of heavy metals produced by the combustion of municipal waste. They also show that there were more dioxins in the waste than were emitted, but that furans were increased at the uncontrolled stack outlet (NITEP 1985).

Dioxin and Furan Emissions

Dioxins and furans are now known to be produced in the lower temperature ranges of boilers and electrostatic precipitators as the result of catalytic reactions involving especially iron and copper. Products of incomplete combustion, especially the carbon matrix containing chlorinated organics, provide sites for production of chlorine, which in the presence of water and oxygen reacts to form dioxins and furans. Good combustion reduces but does not eliminate the organic rings from which dioxins and furans form. Dioxin emissions correlate with (1) particulate quantity, (2) iron and copper content of fly ash, (3) carbon monoxide emissions, and (4) chlorine content of the waste (see Chapter 4). Reducing the stack temperature causes organics such as dioxins to condense out so that they can be collected. A reduction in chlorine content can be expected to reduce dioxin and furan production. This would be especially important if the incinerator was not equipped with an emissions control scrubber (Vogg et al. 1987; Green et al. 1990; Gullett et al. 1989). Recently it has been found that injection of carbon into the flue gases prior to a baghouse is effective in removing dioxins as well as mercury (Teller and Hsieh 1991).

International health experts have concluded that estimates of human health risks resulting from dioxins have been too conservative. *The effect on humans is now believed to be at least 100 times less than has been assumed in the past.* The U.S. EPA has recently substantially revised its standards for dioxins. These revisions essentially remove the concerns about dioxin emissions from good combustion. However, concern about the health effects of heavy metals, especially cadmium, remains.

Heavy Metals Emissions

The metals migrate to either the bottom ash or fly ash. The concentration of metals in the fly ash is related to both the amount of fly ash lifted and the temperature of the flame, which influences the volatilization of metals.

Not all the fine ash produced in combustion of paper and plastics is lifted from the combustion chamber. The amount of particulate matter lifted and the organic and metals content of this particulate is highly sensitive to the combustion conditions, such as the flame temperature, vertical velocities, distribution of underfire and overfire air, and the nature and location of turbulence that takes place. Uncontrolled particulate discharged by starved-air incinerators ranges from 1 to 3.0 lb/t, depending upon the technology and mode of operation.

The percentage of metals in the fly ash is highly dependent not only on the technology, but also on the nature of the waste. Data from tests of a modular (starved-air) incinerator without emission controls burning MSW shows that cadmium was 0.44 percent of uncontrolled particulate sampled from the stack. Cadmium measured from a group of MWIs also employing modular starved-air technology averaged around 0.2 percent of the particulate.

Combustion Temperatures

Maintaining low furnace chamber temperatures substantially reduces the percent of lead and cadmium found in or on the surface of particulate. Starved-air incinerators, which can operate with primary furnace temperatures of 1,400 to 1,600°F, exhibit lower volatile metals emissions than waterwall MSW incinerators, which operate at much higher temperatures.

Figure 5–1 shows the particle size distribution of particulate emissions from an uncontrolled MWI operated at different furnace temperatures. About 60 percent of total particulate was less than one μm in size at the highest temperature (averaging 2,150°F) as compared with 40 percent at 1,640°F and only 15 percent at the lowest temperature tested (about 1,350°F). The difference in fine particulate was attributed to the presence of metals. This research clearly shows the importance of maintaining as low a furnace temperature as possible to minimize the quantity of heavy metals carried by the fly ash. These metals are generally oxides, which react with the chlorine present to form the highly soluble metal chlorides that are of environmental concern (Brady 1991).

5.7 ABSORBING, CONDENSING, AND COLLECTING PARTICULATE

Methods for removing and collecting particulate, organics, and acid gases are described in Chapter 6. In wet scrubbers, acid gases can be absorbed along with solid particulate into water droplets and flushed away. In spray-dryer absorbers, the slurry droplets are dried to powder, which can be collected by the bag filter. Acid gases can also be absorbed on dry powder reagent and collected with the fly ash on a bag filter. An important aspect

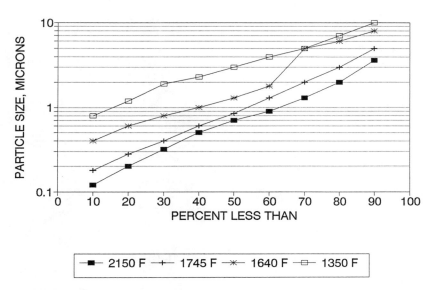

FIGURE 5–1. Particle size distribution of medical waste fly ash at various furnace temperatures. (Adapted from Brady 1991.)

of these processes is condensation, which is highly dependent on temperature.

5.8 EFFECT OF TEMPERATURE ON CONTROL OF EMISSIONS

The collection of acid gases, organic vapors, and volatile metals such as cadmium is substantially affected by the temperature of operation of the emission control device. The California Air Resources Board (CARB) has performed extensive tests of MWIs. Data from these tests have been correlated with scrubbing temperatures. Control of lead was not affected, but cadmium emissions were reduced by a factor of two from 100 to 50 μ/scm as the temperature was reduced from 325°F to about 165°F (Fry 1990).

Dioxins and furans, along with other condensible organics, are effectively controlled as temperatures are reduced. Figure 5–2 shows the effect of emission control temperature on emissions of dioxins and furans, which were reduced 100 times from about 1,000 ng/m^3 to about 10 ng/m^3 as temperatures were reduced from 500°F to 165°F.

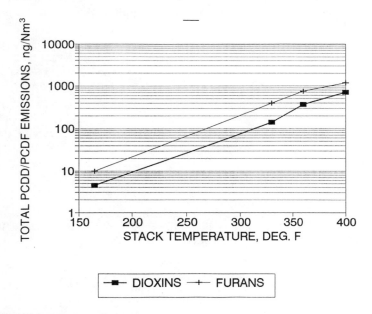

FIGURE 5–2. Effect of emission control temperature on emissions of dioxins and furans from MWIs. (Adapted from Fry et al. 1990.)

Wet scrubbers, which are well suited to small MWIs, operate at temperatures around 145°F, hence are especially effective in controlling acids, organics, and metals by condensation. Fabric filters should not be operated too close to the dew point for fear of causing cementation of the fly ash on the bags. They are thus limited to the range of about 175°F to 200°F, which is still quite effective.

5.9 FLY ASH, SCRUBBER RESIDUES, AND BOTTOM ASH

Due to the volatility of cadmium and lead at combustion temperatures, fly ash can be expected to contain relatively high concentrations of these metals on their surfaces due to condensation as the particles are cooled. Furthermore, due to the presence of HCl in the combustion gases, chlorides of these metals will form. Since these are the most soluble forms of lead, cadmium, and mercury, they are a source of concern if the fly ash and scrubber residues are not treated or disposed of in such a way that these soluble salts do not contaminate the environment and drinking water.

When a wet scrubber is used, the soluble salts from removing acids and from condensing metal salts are absorbed by the scrubbing water and usually discharged to the sewer. Since they become highly diluted after disposal, these salts do not exceed permissible water discharge concentration limits.

When baghouses are used to capture acids and particulate, dry powder fly ash and scrubber residues are produced, which create unique problems of handling and disposal, since they can cause *fugitive dust* and contain the most soluble forms of the toxic heavy metals. Lead and cadmium chlorides are not soluble to a large extent when the pH is greater than 7, except that lead becomes more soluble when the pH exceeds 11 or 12. Excess lime in the scrubber residues can drive the pH up into this range. It is therefore important to manage these residues carefully. They can be treated to convert these metals to other chemical forms, such as carbonates, or they can be bound or encapsulated, such as in a pozzolanic matrix with portland cement.

5.10 EFFECTIVENESS OF EMISSION CONTROLS

Emission controls using alkaline reagents and temperature reduction are extremely effective in controlling acids, organics, and metals, as noted above. Typically, reductions of HCl, organics (including dioxins and fu-

TABLE 5–10. Reduction in Metals Emissions Achieved by Controls (lb/1,000 t)

	Typical Uncontrolled Emissions[a]	Actual Baghouse Test[b]	Actual Scrubber Test[c]
Lead	57.0	0.10	0.03
Iron	7.0	0.025	
Cadmium	6.0	0.005	0.016
Manganese	0.42	0.005	
Chromium	0.36	0.0035	0.030
Arsenic	0.28	0.0000	
Total	71.06	0.1085[c]	0.076
Percent reduction		99.85	99.89

[a] From Morrison 1987; Fry et al. 1990.
[b] Recent test of venturi scrubber (Anderson 2000).
[c] Lead, cadmium, and chromium only, for comparison.

rans), and metals are controlled by more than 99 percent in municipal incinerators and MWIs. Data from two unpublished tests of MWIs are shown in Table 5–10 (Hasselriis 1990, Hasselriis et al. 1991). These high efficiencies are generally confirmed by recent U.S. EPA test results (Durkee and Eddinger 1991). Typical uncontrolled emissions are listed and compared with emissions from one MWI equipped with a baghouse and another equipped with a venturi wet scrubber. There is no question that these controls are highly efficient. The big question is whether or not their installation is justified, especially for small incinerators where such controls can cost far more than a new incinerator.

5.11 RELATING EMISSIONS TO GROUND LEVEL CONCENTRATIONS

Regardless of the absolute values of the concentrations of pollutants in the stack emissions after the emission controls, their significance depends upon how much they become diluted before they reach sensitive receptors. In the process of dispersion, a high degree of dilution can be achieved before the gases reach critical locations, depending on stack height, discharge temperature and velocity, and individual site conditions. Stacks are usually 1.5 to 2.5 times the height of the associated buildings, in order to represent good engineering practice and so that the discharges are not entrained in building downwash.

A dilution factor can be calculated to quantify the change in concentration that takes place as the gases leave the stack and are dispersed in the atmosphere. The dilution factor defined below is based on the point of maximum concentration, found by modeling calculations. It can be calculated for a given facility by the following equations:

$$\text{Dilution factor} = \frac{\text{Concentration in the stack}}{\text{Maximum concentration at the ground}}$$

$$= \frac{1}{(m^3/s) \times [(g/m^3)/(g/s)]}$$

where m^3/s = actual stack flow volume
          = (meters/second) × (square meters stack area)

The actual m^3/s discharged by the stack depends upon the burning capacity of the incinerator, the composition of the fuel, and the type of emission control (if any) used (Hasselriis et al. 1991).

Figure 5–3 shows dilution factors calculated for a wide range of combustion systems, based on studies carried out for CARB (Fry et al. 1990). The stack concentration was calculated from these data and divided by the reported ground level concentrations obtained by modeling. Most of the systems achieved dilution ratios ranging from 20,000 to 500,000; some approached 1 million. The large facilities with high stacks did not necessarily achieve greater dispersion than the small facilities whith short stacks (Hasselriis et al. 1991).

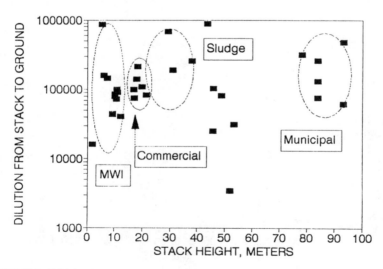

FIGURE 5–3. Dilution factors for various waste combustion systems versus stack height. (From Hasselriis et al. 1991.)

Obviously other site parameters were also important. The variation for most of the MWIs from 40,000 to 120,000 is the result of different stack temperatures, which depend on whether or not waste heat boilers and emission controls were installed.

Effect of Stack Temperature on Dispersion

If the hot gases leaving the combustion chamber are discharged directly to the stack, as is the case with many small incinerators, the buoyancy carries the gases rapidly to a high elevation before horizontal movement disperses the gases. When the gases are cooled by boilers or by water quenching, the buoyancy is greatly reduced. Hence when wet scrubbers or baghouse emission controls are used, dispersion is not as effective.

Figure 5–4 shows ground level concentrations resulting from incinerator discharges of 20 g/s from three basic types, calculated by using the "Screen" model, for a 2,000 lb/h incinerator with a stack height of 30 meters. For the same pollutant discharge rate, the highest local concentration is produced by the most efficient emission control—the wet scrubber with a 145°F discharge temperature (Hasselriis et al. 1991). The lowest concentrations result from the 1,800°F direct discharge. If the effectiveness of the scrubber and the baghouse, which reduces the pollutant dis-

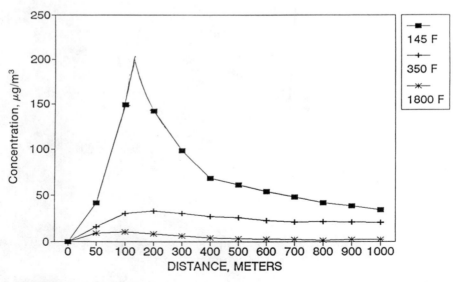

FIGURE 5–4. Ground level concentrations from stack emissions from incinerators with different stack temperatures and emission controls (From Hasselriis et al. 1991.)

TABLE 5–11. Calculation of Ground Level Concentration of Particulate Matter

Waste charging rate	2,000	lb/h
Particulate emissions	0.08	gr/dscf
	= 6.9	lb/h
	= 0.87	g/s
Heat liberated at 7,500 Btu/lb	15	mm Btu/h
Particulate emission factor	0.46	lb/mm Btu
Stack gases at 1,800°F:	37.8	m³/s
Uncontrolled particulate concentration	0.023	g/m³
Dilution factor	20,000	
Ground level concentration	1.15	μg/m³
AAQL (U.S. EPA primary AQS)		
Annual	75	μg/m³
24-h	260	μg/m³
Percent of AAQL	1.5	
Percent of 24-h AAQL	0.44	
Stack gases at 145°F:	11.2	m³/s
Uncontrolled particulate concentration	0.078	g/m³
Controlled particulate at 50% control	0.039	g/m³
Dilution factor	10,000	
Ground level concentration	3.9	μg/m³
AAQL (U.S. EPA primary AQS)		
Annual	75	μg/m³
24-h	260	μg/m³
Percent of AAQL	5	
Percent of 24-h AAQL	1.5	

charge rate, is taken into account, the uncontrolled emissions would result in ground level emissions about three times greater than those from the controlled systems.

Relating Ground Level Concentrations to Health Risk

The ground level concentrations must be compared with acceptable air quality levels established by health authorities. Tables 5–11 to 5–13 show calculations of the ground level concentrations of emissions from a 2,000 lb/h incinerator, making a comparison with typical ambient air quality limits (AAQLs) for particulate, HCl, and cadmium. Note that in accordance with Figure 5–4, a dilution factor of 20,000 is assumed for the 1,800°F stack with no emission control whereas a factor of 10,000 is used with the 145°F stack resulting from a wet scrubber. The scrubber efficiency is also adjusted accordingly for each case.

TABLE 5–12. Calculation of Ground Level Concentration of HCl

Waste charging rate	2,000	lb/h
PVC content at 1% of waste	20	lb/h
HCl at 45% chlorine content	9	lb/h
	= 1.14	g/s
Heat liberated at 7,500 Btu/lb	15	mm Btu/h
HCl emission rate from PVC	0.6	lb/mm Btu
HCl in stack from PVC at 7% O_2	414	ppmv
Stack gases at 1800°F:	37.8	m^3/s
HCl uncontrolled	0.03	g/m^3
Dilution Factor	20,000	
Ground Level Concentration	1.5	$\mu g/m^3$
AAQL	150	$\mu g/m^3$
Percent of AAQL	1	
Stack gases at 145°F:	11.2	m^3/s
HCl uncontrolled concentration	0.10	g/m^3
HCl emitted at 90% control	0.01	g/m^3
Dilution factor	10,000	
Ground level concentration	1.0	$\mu g/m^3$
AAQL	150	$\mu g/m^3$
Percent of AAQL	0.7	

Ground Level Impact of Particulate Emissions

Based on the limit of 0.08 gr/dscf (corrected to 7 percent oxygen), the GLC for particulate matter is 1.5 percent of the AAQL with the hot stack, and 5 percent for the scrubber stack for the annual AAQL based on only 50 percent control efficiency, which results in 0.04 gr/dscf, typical for simple wet scrubbers. It would not seem to be justified to require that particulate emissions be reduced to 0.02 gr/dscf in order to reduce the GLC to 1 percent of the AAQL.

Ground Level Impact of HCl Emissions

Table 5–12 shows that HCl concentrations would also be only about 1 percent of the AAQL. Note that the HCl emissions are 9 lb/h in this example for a 2,000 lb/h charging capacity. The emissions could be reduced to the required 4 lb/h by reducing the chlorine in the waste or by reducing the capacity of the incinerator to 880 lb/h. If the state required HCl to be reduced to 100 ppm, the capacity of the incinerator could be reduced to about 480 lb/h, unless the chlorine in the waste can be reduced.

TABLE 5-13. Calculation of Ground Level Concentration of Cadmium

Waste charging rate	2,000	lb/h
Cadmium emissions	1.0	ppm
	0.002	lb/h
	0.00025	g/s
Heat liberated at 7,5000 Btu/lb	15	mm Btu/h
Cadmium emission factor	0.00014	lb/mm Btu
Stack gases at 1,800°F:	37.8	m^3/s
Uncontrolled cadmium concentration	6.6	$\mu g/m^3$
Dilution factor	20,000	
Ground level concentration	0.33	ng/m^3
AAQL (New York State Air Guide)	0.56	ng/m^3
Percent of AAQL	60	
Stack gases at 145°F:	11.2	m^3/s
Actual cadmium stack concentration	1.95	g/m^3
Controlled cadmium at 90% control	0.2	$\mu g/m^3$
Dilution factor	10,000	
Ground level concentration	0.02	$\mu g/m^3$
AAQL	0.56	$\mu g/m^3$
Percent of AAQL	4	

Ground Level Impact of Cadmium Emissions

Assuming cadmium emissions are 1 pound per million pounds of waste, typical of MWIs, the uncontrolled emissions of the hot stack are a relatively high 60 percent of the AAQL. However, as seen in Table 5–7, Cedars Sinai was found to emit 2.2 ppm of cadmium, which would raise the GLC above the AAQL. In this situation, reducing the cadmium in the waste might be considered to avoid having to install a scrubber. With the scrubber operating at conservative 90 percent control, the GLC is reduced to 4 percent of the AAQL. Since 99 percent control is likely, this would probably be reduced to less than 0.5 percent.

Health Risk from Cadmium

The CARB performed an elaborate analysis of the health risk derived from cadmium emissions from MWIs. The maximum annual ground level concentration was estimated for a typical incinerator *without emission controls,* having a feed rate of 783 lb/h (St. Agnes). The health risk estimated for this MWI was estimated to be 1.5 to 9.2 additional cancer cases for the maximally exposed individual by inhalation, based on a 1 g/s emission rate. The measured emission rate of this MWI was only 0.2 g/s, so these estimates must be reduced to 0.4 to 2.0 additional cases. The CARB report also estimated the total multipath health risk to be 7.4 to 45 additional cases, which corrected to 0.2 g/s become 1.5 to 9 additional cases.

Health risks from 1 to 10 in 1 million are generally treated as not needing action, whereas those less than 1 in a million are considered to have no significance at all. If the cadmium in the waste were reduced by a factor of 4, the health risk would be reduced to about 0.5 to 2 additional cases. It appears that in this particular example, reduction of cadmium content would result in lowering the risk from no concern to no significance.

It cannot be assumed that these numbers would apply to all situations. Especially where there are high-rise buildings in the vicinity of an incinerator stack, detailed modeling must be carried out in order to comply with the requirements of many states. If higher risk numbers were estimated, justification for removing cadmium from the waste or installing a scrubber might well be found.

Comparison of Alternatives

As can be seen from the above discussions and examples, there are many options that may be exercised in order to achieve acceptable environmental impacts from combustion of medical and other wastes. Undesirable toxics may be avoided as components of the waste by purchasing disciplines, or by other means of removal. Good combustion practices, such as avoiding excessive furnace temperatures, can prevent volatilization of metals into the gas stream. Emission control devices can remove almost all pollutants of concern, and finally, if the optimum stack height and location are provided, sufficient dispersion and dilution can be achieved to bring ground level and critical elevated concentrations safely below those acceptable from a health standpoint.

A more detailed study of these factors is illustrated in Tables 5–14, 5–15, and 5–16. These tables trace HCl, particulate, metals, and toxic equivalent 2,3,7,8-TCDD from the waste to the ground level concentrations, arriving at the percent of acceptable GLC for three dilution factor (DF) values: 500, 10,000, and 100,000.

Table 5–14 shows the results of this analysis for a MWI with heat recovery boiler without emission controls, discharging gases to the stack at 450°F. The partition factor shows the fraction of the metals in the waste that enters the gas stream. These factors and actual uncontrolled stack concentrations were measured during U.S. EPA tests of a MWI, and were used to calculate the actual stack concentration at 450°F (Durkee and Eddinger 1991). The acceptable GLC values were abstracted from a New

TABLE 5–14. MWI with Heat Recovery Boiler, No Emission Controls

Pollutant	lb/t in Waste	Removal Factor	Partition Factor	Emission Control Factor	Stack[f] Conc. (μg/m³ at 450°F)	Acceptable GLC (μg/m³)	GLC (DF = 500)	% of AGLC (DF = 10,000)	Annual Average (DF = 100,000)
MCl[b]	8.16	1	1	1	967,611	7	922%	46.08%	4.61%
Particulate[c]	0.86	1	1	1	101,984	50	14%	0.68%	0.07%
Arsenic	0.0005	1	0.12	1	7	0.00023	214%	10.68%	1.07%
Cadmium	0.0013	1	0.77	1	117	0.00056	1,398%	69.89%	6.99%
Chromium	0.0020	1	0.05	1	12	1.2	0.07%	0.00%	0.00%
Chrome VI[d]	0.0002	1	0.05	1	1	8E-05	98%	4.89%	0.49%
Lead	0.0268	1	0.45	1	1,428	1.5	6%	0.32%	0.03%
Mercury	0.0048	1	1	1	571	0.12	32%	1.59%	0.16%
Total DD/DF	1.4E-06[a]	1	1	1	0				
T.E. TCDD[e]	4.8E-08	1	1	1	5.7E-03	3E-08	1260%	62.98%	6.30%

[a] $1\ \mu g/m^3 \times 6.93E\text{-}6 = 6.93\ E\text{-}6\ lb/h$ [corrected to 7% O_2].
[b] HCl averaged 1,627 ppm (2,395 mg/m³) during tests.
[c] Particulate averaged 0.11 gr/dscf (0.25 g/m³) during tests.
[d] Chrome VI is assumed to be 10% of chromium.
[e] Toxic Equivalent (T.E.) TCDD is assumed to be 1/30 of total DD/DF.
[f] Combustion dilution factor for boiler *only* at 450°F is 1/247 = 0.405.

TABLE 5–15. MWI with Boiler, No Emission Controls, Remove Toxics

Pollutant	lb/t in Waste	Removal Factor	Partition Factor	Emission Control Factor	Stack[f] Conc. ($\mu g/m^3$ at 450°F)	Acceptable GLC ($\mu g/m^3$)	GLC (DF = 500)	% of AGLC (DF = 10,000)	Annual Average (DF = 100,000)
HCl[b]	40.81	0.2	1	1	193522	7	184%	9.22%	0.92%
Particulate[c]	0.86	1	1	1	101984	50	14%	0.68%	0.07%
Arsenic	0.0005	1	0.12	1	7	0.00023	214%	10.68%	1.07%
Cadmium	0.0064	0.2	0.77	1	23	0.00056	280%	13.98%	1.40%
Chromium	0.0020	1	0.05	1	12	1.2	0.07%	0.00%	0.00%
Chrome VI[d]	0.0002	1	0.05	1	1	8E-05	98%	4.89%	0.49%
Lead	0.0268	1	0.45	1	1428	1.5	6%	0.32%	0.03%
Mercury	0.0048	1	1	1	571	0.12	32%	1.59%	0.16%
T.E. TCDD[e]	2.4E-07[a]	0.2	1	1	1.1E-03	3E-08	252%	12.60%	1.26%

[a] 1 $\mu g/m^3$ × 6.93E-6 = 6.93 E-6 lb/h [corrected to 7% O_2].
[b] HCl averaged 1,627 ppm (2,395 mg/m^3) during tests.
[c] Particulate averaged 0.11 gr/dscf (0.25 g/m^3) during tests.
[d] Chrome VI is assumed to be 10% of chromium.
[e] Toxic Equivalent (T.E.) TCDD is assumed to be 1/30 of total DD/DF.
[f] Combustion dilution factor for boiler *only* at 450°F is 1/2.47 = 0.405.

TABLE 5-16. MWI with Boiler and Fabric Filter Emission Controls

Pollutant	lb/t in Waste	Removal Factor	Partition Factor	Emission Control Eff'y.	Emission Control Factor	Stack[f] Conc. ($\mu g/m^3$ at 300°F)	Acceptable GLC ($\mu g/m^3$)	GLC (DF = 500)	% of AGLC (DF = 10,000)	Annual Average (DF = 100,000)
HCl[b]	8.16	1	1	96	0.04	44341	7	42%	2.11%	0.21%
Particulate[c]	0.86	1	1	92	0.08	9347	50	1%	0.06%	0.01%
Arsenic	0.0005	1	0.12	69	0.31	3	0.00023	76%	3.79%	0.38%
Cadmium	0.0013	1	0.77	99	0.01	1	0.00056	16%	0.80%	0.08%
Chromium	0.0020	1	0.05	79	0.21	3	1.2	0.02%	0.00%	0.00%
Chrome VI[d]	0.0002	1	0.05	79	0.21	0	8E-05	24%	1.18%	0.12%
Lead	0.0268	1	0.45	99	0.01	16	1.5	0%	0.00%	0.00%
Mercury	0.0048	1	0.45	0	1	654	0.12	36%	1.82%	0.18%
T.E. TCDD[e]	4.8E-08[a]	1	1	95	0.05	0	3E-08	72%	3.61	0.36%

[a] 1 $\mu g/m^e \times 6.93E\text{-}6 = 6.93$ E-6 lb/h (corrected to 7% O_2).
[b] HCl averaged 1.627 ppm (2.395 mg/m³) during tests.
[c] Particulate averaged 0.11 gr/dscf (0.25 g/m³) during tests.
[d] Chrome VI is assumed to be 10% of chromium.
[e] Toxic Equivalent (T.E.) TCDD is assumed to be 1/30 of total DD/DF.
[f] Combustion dilution factor for boiler plus FF at 300°F is 1/2.156 = 0.464.

York state permit application (Hasselriis 1992). A DF of 500 represents an unfavorable site condition where gases from a short stack are intrained by the downwash of a nearby building. Evidently the AGLC is exceeded for many pollutants. However, with a higher stack and/or smaller or more distant nearby buildings corresponding to a dilution of 10,000, the average GLC falls below the AGLC. With a DF of 100,000, which is likely with an adequate stack height and no large buildings nearby, no GLC exceeds 10 percent of the AGLC.

Table 5–15 shows the same MWI system, assuming that 80 percent removal (0.2 removal factor) of HCl, cadmium, and TCDD could be achieved. The GLCs at a DF of 10,000 are now closer to 10 percent of the AGLC.

The effect of adding a fabric filter emission control to the MWI with boiler is shown in Table 5–16. Emission control efficiencies found by the EPA study are included in the calculations (Durkee and Eddinger 1991). The GLCs resulting from a DF of 500 are now less than 76 percent of the AGLC, and less than 4 percent for a DF of 10,000.

These examples show that the emissions from a MWI without emission controls may be acceptable, and that when the emissions are marginal, they can be significantly reduced by preventing potential toxic pollutants from entering the waste stream to be burned. They also show that increasing the stack height may make a MWI without emission controls perform as well as one with less favorable stack and building configuration.

In conclusion, we see that when site conditions are unfavorable, producing a DF as low as 500, the provision of emission controls may be necessary in order to achieve acceptable GLCs comparable to those with favorable dispersion conditions, but lacking emission controls. The cost of a higher stack may have to be weighed against the cost of adding emission controls.

5.12 SUMMARY

Many factors determine how potential pollutants in the waste will enter the atmosphere and how much they will affect ground level concentrations and perhaps become environmental and health concerns. Some potential pollutants such as highly chlorinated plastics and solvents can be avoided by not purchasing materials containing them in the first place. Other pollutants, such as lead and cadmium, could be avoided by restricting their use as fillers and colorants in plastics, textiles, and paper.

Practically all the particulate, acid gases, heavy metals, and organics emitted from the combustion process can be removed by postcombustion emission controls using condensation and absorption so that they can be removed by a wet scrubber, baghouse, or other emission control device. Low temperatures (below 250°F) are extremely effective in condensing these pollutants, including the relatively difficult-to-remove sulfur dioxide. Lime injection with a fabric filter is effective in removing the metals and acids, and recently it has been found that carbon injection succeeds in removing mercury and even the dioxins. However, it may not be justified to apply postcombustion controls when the ground level concentrations after dispersion are only a small percentage of concentrations considered acceptable by regulators and health authorities.

It is important to model emissions at specific sites since unfavorable factors may result from low stack temperatures, low stacks, and proximity to adjacent buildings.

The examples cited above indicate that small on-site MWIs *without emission controls* may not represent the health risks generally believed to be present. When estimated ground level concentrations are higher than a small percentage of AAQLs and the purchase of a scrubber is indicated, it may be possible to select the alternative of cleaning the waste by avoiding the purchase of items and materials containing heavy metals, such as cadmium, and/or chlorine-bearing plastics, in the waste. The ash residues from incineration of medical wastes contain soluble salts and heavy metals that are prone to leach. Reduction of chlorides and metals in the waste will reduce the quantity of leachable salts.

REFERENCES

Bergvall, G. and J. Hult. 1986. "Technology, economics, and environmental effects of solid waste treatment." DRAV No. 33, Statens naturvårdsverk, Solna, Sweden.

Brady, J. D. "Recent developments in pollution control systems for chemical and infectious waste incinerators." Mid-West American Institute of Chemical Engineers Meeting, St. Louis, MO, February, 1991.

Brown, H. L. "Thomas Jefferson University Hospital Waste Characterization Study." Drexel University, Philadelphia, 1989.

Corbus, D. "A comparison of air pollution control equipment for hospital waste incinerators." Paper 90-27.4 presented at the 83rd Air and Waste Management Association Meeting, Pittsburgh, June 1990.

Durkee, K. R. and J. A. Eddinger, "Status of U.S. EPA regulatory program for medical waste incinerators—test program and characterization of emissions," Incinerator Conference on Thermal Treatment of Radioactive, Hazardous, Chemical, Mixed, and Medical Wastes, Knoxville, TN, May 13–17, 1991.

Fry, B., et al. "Technical support document to proposed dioxins and cadmium control measure for medical waste incinerators." California Air Resources Board, Sacramento, CA, May, 1990.

Green, A. et al. "Toxic products from co-combustion of institutional waste." Paper 90-38.3 presented at the 83rd Annual Air and Waste Management Association Meeting, Pittsburgh, June, 1990.

Gullett, B. K., K. R. Bruce, and L. O. Beach. "Formation mechanisms of chlorinated organics and impacts of sorbent injection." *International Conference on Municipal Waste Combustion,* Hollywood FL, April 1989.

Hasselriis, F. 1984. *Refuse-derived fuel processing,* Stoneham, MA, Butterworth.

Hasselriis, F. "Minimizing refuse combustion emissions by combustion control, alkaline reagents, condensation, and particulate removal." Synergy/Power Symposium on Energy from Solid Wastes, Washington, DC, October 1986.

Hasselriis, F. 1987. "Optimization of combustion conditions to minimize dioxin emissions." *Waste Manage. & Res.* 5 (3):311–326.

Hasselriis, F. "Relationship between waste composition and environmental impact." Paper 90-38.2 presented at the 83rd Annual Meeting of the Air and Waste Management Association, Pittsburgh, June 1990.

Hasselriis, F., D. Corbus, and R. Kasinathan. "Environmental and health risk analysis of medical waste incinerators employing state-of-the-art emission controls." Paper No. 91-30.3 presented at the 84th Annual Meeting of the Air and Waste Management Association, Vancouver, Canada, June 1991.

Hasselriis, F. "Relationships between variability of emissions and residues from combustion of municipal solid waste and waste composition and combustion parameters." Prepared for the American Society of Mechanical Engineers Solid Waste Processing Conference, Detroit, MI, May 1992.

Hasselriis, F. and R. Kasinathan. "Effect of waste stream modification on impact of hospital waste incinerators on ground level and flagpole receptor concentrations as compared with acceptable risk factors, based on six permit applications." Paper 92-23.03 prepared for the 85th Annual Meeting of the Air and Waste Management Association, Kansas City, MO, June 1992.

Hay, D. J., A. Finkelstein, and R. Klicius. "The National incinerator testing and evaluation program: An assessment of a) two-stage incineration and b) pilot-scale emission control." Air Pollution Control Association Annual Meeting, Minneapolis, MN, June 1986.

Hickman, D. C., D. P. Y. Chang, and H. Glasser. "Cadmium and lead in bio-medical waste incinerators." Air & Waste Management Association Conference, Pittsburgh, June 1989.

Kerstetter, J. and J. K. Lyons. "Heavy metal concentration in selected paper samples." Washington State Energy Office, Olympia, WA, June 1989.

Morrison, R. "Hospital waste combustion study—data gathering phase." U.S. EPA Contract No. 68–02–4330, Ottawa, Ontario 1987.

NITEP. "Two-stage combustion (Prince Edward Island)." The National Incinerator Testing and Evaluation Program. Environment Canada, Report EPS 3/UP/1, Ottawa, Ontario, 1985.

NITEP. "Air pollution control technology." The National Incinerator Testing and Evaluation Program. Environment Canada, Report EPS 3/UP/2, Ottawa, Ontario, September, 1986.

NITEP. "Environmental characterization of mass burning incinerator technology at Quebec City." The National Incinerator Testing and Evaluation Program. Environment Canada, Report EPS 3/UP/5, Ottawa, Ontario, June 1988.

Teller, A. and J. Y. Hsieh, "Control of Hospital Waste Incineration Emissions—Case Study." Paper No. 91-30.5 presented at the 84th Annual Meeting of the Air and Waste Management Association, Vancouver, Canada, June 1991.

U.S. EPA. "Municipal waste combustion study: Emission data base for municipal waste combustors." Office of Solid Waste, EPA/530-SW-87-021b, Washington, DC, June 1987.

U.S. EPA. "Characterization of products containing lead and cadmium in municipal solid waste in the United States 1970 to 2000." EPA/530-SW-89-015C, U.S. EPA Office of Solid Waste, Washington, DC, 1989.

Vogg, H., M. Metz, and L. Steiglitz. 1987. "Recent findings on the formation and decomposition of PCDD/PCDF in municipal solid waste incineration." *Waste Manage. Res.*, 5(3): 285–294.

6

Postcombustion Cleanup

David Corbus

6.1 INTRODUCTION

Postcombustion cleanup refers to processes occuring downstream of the combustion chamber that reduce pollutant emissions. Postcombustion cleanup for medical waste incinerators typically consists of air pollution control (APC) equipment, which removes pollutants from the exhaust stream. Increased state regulations for the incineration of medical wastes require new and existing medical waste incinerators to meet strict pollutant emission limitations. These regulations do not favor pollution prevention as a method of reducing pollutant emissions. In most cases, expensive APC equipment is required to meet these new regulations. Separation of chlorine- and metal-bearing wastes from the incinerator waste stream prior to incineration, or material substitution for these wastes, could result in environmentally acceptable pollutant emissions without the need for APC equipment in some cases. This approach would also negate the problem of having to dispose of a potentially toxic fly ash or scrubber effluent that results from removing toxic pollutants from the exhaust gas.

It is important to have a good understanding of the current APC equipment used on medical waste incineration facilities in order to evaluate the emissions control needed for facilities practicing pollution prevention. The most common types of APC equipment being installed on medical waste incinerators to meet new state regulations are described in the following sections. The need for APC systems for medical waste incineration facilities practicing pollution prevention is then discussed, and recommendations are given for postcombustion cleanup for incineration facilities using pollution-prevention procedures.

128

6.2 DESCRIPTION OF AIR POLLUTION CONTROL SYSTEMS

The trend in state regulations for new medical waste incinerators is to set emission limits for particulates at either 0.03 grains per dry standard cubic foot (gr/dscf) or 0.015 gr/dcsf. State regulations for HCl emission limits usually require a 90 percent reduction in HCl by the air pollution control system, or a concentration of 30 ppm in the stack gas. These emission limits are considered in the description of APC equipment described below. Regulations for existing medical waste incinerators, or for new small units (typically less than 200 lb/h), are usually more liberal and may have significantly higher emission limits.

There are many different kinds of air pollution control systems. Systems include settling chambers, mechanical cyclones, electrostatic precipitators, and a variety of wet scrubbers and dry scrubbers. Settling chambers reduce the velocity of the gases, thereby permitting the larger particles to settle out. Mechanical cyclones rely on centrifugal force to separate particles from the flue gas stream. Neither system is able to meet the new stricter emission regulations being promulgated for medical waste incinerators. Electrostatic precipitators induce an electrical charge to particles in the flue gas, which are then attracted to plates that have an opposite charge. Particles are then collected on the plates and removed from the incinerator flue gas. The high cost of electrostatic precipitators is one of the main reasons these systems are not found on medical waste incinerators (MWIs), which are smaller in size than municipal solid waste incinerators (MSWIs) where electrostatic precipitators have been used successfully. Additional reasons for electrostatic precipitators not being used on MWIs include problems with corrosion, fouling, and the passing of large particles through the system. Wet scrubbers and dry scrubbers (baghouses with alkaline injection upstream) are the two APC systems that have been used successfully on medical waste incinerators to meet new state emission regulations. These systems can control organic pollutants by condensation and absorption; particulates and metals by dust collection or impaction; and acids by reaction with alkaline materials (lime, soda ash, and others).

Wet Scrubbers

Wet scrubbers are the most common APC equipment used on MWIs today. The main reason for this is that they have been the least costly and the easiest to operate for this application. In addition, most older MWIs were required to have only acid gas removal and not high particulate removal efficiencies, so scrubbers did not have to be high energy venturi-type scrubbers. Wet scrubbers may be used for removal of acid gases or both acid gases and particulates.

Venturi, packed-bed, and spray towers are the most common types of wet scrubber sytems used on MWIs. Venturi scrubbers are used primarily for particulate removal, and packed-bed scrubbers are used primarily for acid gas removal. The two may be used together for effective control of both acid gases and particulates. Spray towers may be used for both particulate and acid gas removal, but they are not effective enough to meet the new hospital incinerator regulations for particulates, hence they are primarily used in series with a venturi scrubber. Impingement tray scrubbers are another type of scrubber, which may be used for particulate and acid gas removal, but like the spray tower scrubber, their particulate removal efficiencies cannot meet new regulations. Packed-bed and spray towers used with venturi scrubbers are more common than impingement tray scrubbers.

Other types of scrubbers with potential applications for MWIs include the collision scrubber, the ejector scrubber, and the wet ionizing scrubber. These scrubbers all rely on either impaction for particulate capture or absorption for gas removal.

Wet scrubbers use large liquid droplets to capture relatively small dust particles. The droplets collect particles by using a process called *impaction*. Particles impact onto droplets that can be removed from the system. Acid gases are removed in a scrubber by dissolving the gaseous pollutants in a liquid, a process known as *absorption*. To remove a gaseous pollutant by absorption, the exhaust stream must be passed through (brought in contact with) a liquid. To achieve good acid gas absorption, wet scrubbers must provide a large interfacial contact area between the gas and liquid phases, provide good mixing of the gas and liquid phases (turbulence), and allow sufficient contact time for the absorption to occur. The processes of impaction and absorption for a venturi scrubber are described below. This type of scrubber is the most common type of wet scrubber being installed on medical waste incinerators today.

Venturi scrubbers rely on impaction to achieve particulate removal, as opposed to filtration as in baghouse technology. A venturi scrubber consists of a liquid sprayed upstream from a vessel containing a converging and diverging cross-sectional area. The portion of the venturi that has the minimal cross-sectional area and consequently the maximum gas velocity is commonly referred to as the *throat*. The throat can be circular or rectangular. As the gas stream approaches the venturi throat, the gas velocity and turbulence increase. Liquid droplets serve as the collection media and can be created by two different methods. The most common method is to allow the shearing action of the high gas velocity in the throat to atomize the liquid into droplets. The other method is to use spray nozzles to atomize the liquid by supplying high pressure liquid through small orifices.

To attain a high collection efficiency, venturi scrubbers must achieve gas velocities in the throat in the range of 10,000 to 40,000 ft/min. These high gas velocities atomize the water droplets and create the relative velocity differential between the gas and the droplets to effect particle-droplet collision. The effectiveness of a venturi scrubber is related to the square of the particle diameter and the difference in velocities of the liquid droplets and the particles.

From the venturi section the gas enters a large chamber for separation of particles and then passes through a packed tower or spray tower scrubber for removal of acid gases. A quench section is included in the scrubber system to bring the temperature of the incinerator flue gas to the saturation temperature. A waste heat recovery boiler may be located upstream of the quench section for heat recovery and to help reduce flue gas temperatures.

The design gas velocity in the venturi throat depends upon a variety of parameters, including the required particulate removal efficiency and the particulate matter size distribution. The removal efficiency falls off rapidly for small particulates. Venturi scrubber particulate collection efficiency is generally correlated with the pressure drop across the venturi, as opposed to throat velocity. The pressure drop is easy to measure and has a direct impact on the size of the induced draft fan required. Fan size has a direct impact on the electrical operating costs for the system. The required pressure drop across a standard venturi scrubber increases exponentially as particulate removal requirements increase. The capture of large particulates is easily accomplished. Higher particulate collection requirements, such as those required to achieve 0.03 or 0.015 gr/dscf, require collection of smaller particulates, which are more difficult to capture.

Acid gas capture begins in the venturi scrubber and is completed in the absorption tower. The absorption tower may consist of either a packed bed or a spray tower configuration. The liquid is sprayed from the top and flows downward across the bed. Typically, sodium hydroxide (NaOH) is used with water to neutralize the absorbed acid gases in a packed-bed scrubber.

Particulate matter is captured by the caustic water solution injected into the venturi scrubber. This particulate filled water becomes the scrubber liquid at the base of the acid gas absorber and is mixed with the fluid sprayed into the scrubber for acid gas control. A significant part of the particulate loading from standard venturi scrubbers can be from insoluble material contained in water droplets that escape from the absorption tower. Much of this material consists of insoluble salts that have no adverse environmental impact but still contribute to outlet particulate loadings.

Venturi scrubbers have had good success in the field and can easily meet the new regulatory requirements of 90 percent reduction for HCl emissions. Systems have been proven in the field to achieve particulate levels of 0.03 gr/dscf on medical waste incinerators, and vendors are currently installing them under guarantees to meet emission limits of 0.015 gr/dscf. The major components of a commercial venturi packed-bed scrubber system are as follows:

- Venturi with variable throat section and spray nozzles
- Integral quench system or chamber for quenching flue gases from the incinerator to the adiabatic saturation temperature upstream of the venturi scrubber throat
- Packed tower section comprising packing media, a subcooler section, and an outlet mist eliminator (demister) to remove entrained water droplets
- Induced draft fan
- Water circuitry including recirculation system with pumps, piping valves, and fittings
- A pH neutralization system with a caustic (sodium hydroxide) feed tank, pump set, piping and controllers
- System controls and instrumentation

Figure 6–1 shows the components of a venturi scrubbing system.

Dry Scrubbers

Dry scrubber is a broad term that has been used to describe any APC equipment consisting of dry alkaline injection upstream of a particulate removal device. Hence, dry scrubbers may consist of lime injected upstream of an electrostatic precipitator or a baghouse. As mentioned earlier, electrostatic precipitators are not commonly used on medical waste incinerator facilities, hence the discussion on dry scrubbers is limited to alkaline injection upstream of a fabric filter (baghouse).

Fabric filtration is one of the most common techniques used to collect particulate matter. A fabric filter is a collection of bags constructed of a fabric material (nylon, wool, or other) hung inside a housing. The combustion gases are drawn into the housing, pass through the bags, and are exhausted from the housing through a stack to the atmosphere. When the exhaust stream from the incinerator is drawn through the fabric, the particles are retained on the fabric material, while the cleaned gas passes

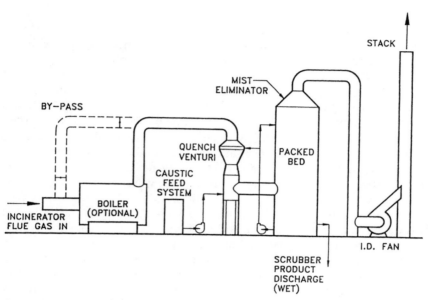

FIGURE 6-1. Venturi scrubber/packed tower system.

through the material. The collected particles are then removed from the filter by a cleaning mechanism, typically by using blasts of air. The removed particles are stored in a collection hopper until they are disposed of, and they are referred to as *fly ash*.

With a new filter, the open areas in the fabric are of sufficient size that particles easily penetrate the bag. Over time, a cake builds on the bag surface, and this cake acts as the primary collection medium. The operating temperature of the fabric filter is of critical importance. The exhaust gas from a hospital incinerator facility that incinerates chlorine wastes contains HCl, hence the unit must be operated at sufficiently high temperatures to ensure that no surfaces drop below the acid dew point. Otherwise, condensation of HCl will result in corrosion of the housing or bags. The boiling point of HCl (aqueous hydrochloric acid) is 110°C (230°F); gas temperatures should be maintained at 150°C (300°F) to ensure that no surfaces are cooled below the dew point. Above a maximum temperature that is dependent on filter type, bags will degrade or in some cases fail completely. Gas temperatures must be kept safely below the allowed maximum.

Fabric filters by themselves do not result in any removal of acid gases, hence they are not used on medical waste incinerators without additional processes for acid gas removal. In order to remove acid gases along with particulates, an alkaline absorbent may be injected upstream of a fabric filter. In addition to removing acid gases, the alkaline absorbent also helps prevent bags from corrosion by HCl (because it neutralizes the HCl). These systems are generally referred to as dry scrubbers.

Dry scrubber/fabric filter systems for acid gas and particulate removal are of two types: spray dryer/fabric filters and dry injection/fabric filters. Both systems introduce an alkaline absorbent into the flue gas upstream of a fabric filter. The fabric filter then removes the acid gases absorbed onto the alkaline particulates as well as other particulates. The main difference between a spray dryer and a dry injection system is the method of introducing the alkaline absorbent. Spray dryer systems use a lime slaker system whereby the lime is mixed with water to form a paste and then fed into a spray reactor. Since the lime is in a slurry form, this type of system is sometimes called a *wet/dry system,* since the absorbent is introduced into the system in a wet state and removed from the system after it has been dried from the heat of the gases. Injection of lime slurry into the spray reactor can be through dual fluid nozzles where air is used to atomize the slurry, or through rotary atomizers. The acid gases react with the slaked lime in the slurry droplet and water is then evaporated by sensible heat from the gas stream until only the solid reaction products are left for collection at the baghouse.

In a dry injection system, the alkaline absorbent (lime) is injected directly into the gas stream where it absorbs acid gases. More alkaline absorbent per pound of HCl removal is required for a dry injection system than for a spray dryer, and the reaction is not as efficient. Yet these systems have demonstrated their effectiveness. Since there is no lime slaker preparation and no spray reactor, these systems are less expensive than a spray dryer. Spray dryers are found widespread in MSWIs, but are generally too expensive for smaller applications such as medical waste incinerators. To date, there are no known installations of a spray dryer on a medical waste incineration facility. Dry injection systems are becoming increasingly popular where very high particulate removal efficiencies are required along with acid gas scrubbing. These systems have been success-fully installed at medical waste incineration facilities. Figure 6–2 shows a schematic for a dry injection/fabric filter system. The major components of a dry injection/fabric filter system are as follows:

- Dry additive feed system comprising dry reagent silo and weighing system
- Waste heat recovery boiler to reduce the temperature of the flue gases and for energy recovery (this is optional, gases may be reduced in temperature without a boiler)

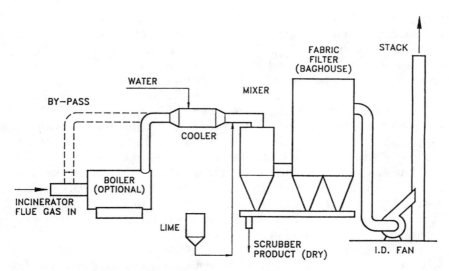

FIGURE 6–2. Dry sorbent injection/fabric filter system.

- Heat exchanger for additional flue gas temperature control
- Dry reactor for mixing of lime with flue gases (some systems just comprise a special nozzle with injection directly into the breeching)
- Fabric filter (baghouse) for high efficiency removal of particulates
- Induced draft fan
- System controls and instrumentation

Emissions

A dry injection system can achieve particulate emission levels below 0.01 gr/dscf. Venturi scrubbers have demonstrated 0.03 gr/dscf in the field, and vendors are currently guaranteeing 0.015 gr/dscf, although no test results are available from these facilities.

Ash and Liquid Effluent Disposal

The waste left over from the APC system can have an important effect on costs, operations, and environmental impacts. Wet scrubber systems produce a liquid effluent as waste, which is usually discharged directly to the sewer system. Baghouse systems produce fly ash, which is classified as either a solid waste, a special handling waste, or a hazardous waste. It is disposed of in an appropriate landfill according to its designation. The effluent from a venturi scrubber system from a medical waste incinerator may contain concentrations of heavy metals in excess of local sewer pretreatment standards (which can be very strict) or federal regulations.

At hospitals the scrubber effluent is usually mixed with other liquid discharges from the hospital before being discharged to the hospital sewer, providing a safe dilution factor for the effluent to ensure compliance with all regulatory requirements. In the case of direct discharge to the sewer system under strict sewer pretreatment standards, a specially designed spray evaporator may be employed in series with the wet scrubber to dry the liquid waste for disposal in a landfill. A more common dilemma is how to classify fly ash from the baghouse. If this waste is categorized as hazardous, the annual cost of the system can increase significantly. This can be an important drawback for dry injection systems. The general trend seems to be to classify this fly ash as a special handling waste, which puts it between a solid waste and a hazardous waste in terms of disposal costs.

Another option is to use a dry injection/fabric filter system with a wet scrubber downstream. This type of system will capture trace metals in the baghouse fly ash instead of in the scrubber effluent. The fly ash must then be disposed of in a safe manner. Additional acid gas and particulate removal is achieved by the wet scrubber. These types of systems have been installed successfully on medical waste incinerators, but the redundancy of having a wet and a dry scrubber is generally not required, and these systems are more costly then venturi scrubbers or dry injection/fabric filter systems.

6.3 EFFECT OF POLLUTION PREVENTION ON POSTCOMBUSTION OPERATIONS

Pollution prevention can have a strong effect on reducing pollutant emission concentrations, and hence can be an effective process for postcombustion control. For some medical waste incinerators, it makes sense to consider pollution prevention as the method to control pollutant emissions, because it can eliminate the high capital costs for APC equipment and it does not result in the creation of a potentially toxic fly ash or liquid effluent. When does it make sense to favor pollution prevention over installation of an APC system? If pollution prevention can achieve pollutant emission limits that result in acceptable environmental impacts, it should be favored over installation of APC equipment.

Starved air incinerators are capable of meeting emission limits of 0.08 g/dscf, without any APC equipment. In order for them to meet this standard, they must be operated with less than stoichiometric air in the primary chamber and excess air in the secondary chamber. This results in less entrainment of particulates because less turbulence is created. If there is no chlorine in the waste being incinerated, a starved air incinerator will produce no HCl. Emissions of sulfur dioxide, the other acid gas often regulated by state medical waste regulations, are usually lower than typical emission requirements before APC controls (if not, pollution prevention procedures can also be used to control this acid gas). This means that a properly operated starved air incinerator practicing pollution prevention procedures could meet the emission limits for acid gases without any APC system, and it could meet reasonable particulate emission limits (0.08 g/dscf). The objective of wet and dry scrubbers is to remove particulates and acid gases and the toxic metals and dioxins that condense on the particulates.

High particulate removals are required of incinerators burning wastes containing toxic metals, because the metals must be removed from the exhaust gas by condensing them onto particulates. In a similar manner, dioxins produced from chlorine in the waste stream must be removed by condensing them onto particulates. Pollution prevention resulting in the reduction of toxic metals and chlorine in the waste stream would eliminate the need for the extremely high particulate emission limits required by most new state regulations for medical waste incinerators (typically 0.03 gr/dscf), because the particulates would contain little or no toxic metals or dioxins.

No pollution prevention program is 100 percent effective in disallowing wastes that contain some toxic metals, hence large incinerators (greater than 2,000 lb/h) would need to consider APC equipment. The APC equipment for these incinerators should not have to meet the strict particulate emission limits of 0.015 gr/dscf (required by many states for incinerators with capacities greater than 2,000 lb/h) because of the reasons discussed above. High temperature baghouses could be considered for particulate control for large incinerators with pollution prevention controls. High temperature baghouses can withstand temperatures of up to 1,200 to 1,400°F. Flue gases would still need to be cooled upstream of a high temperature baghouse by a waste heat recovery boiler or other heat exchanger (heat recovery is often economical for large incinerators).

In general, when incinerators include pollution prevention controls and have capacities less than 500 lb/h, installation of APC equipment should not be required. When starved air incinerators with capacities between 500 and 2,000 lb/hr include pollutant prevention controls, an evaluation, based upon an air quality modeling analysis described in Section 6.4, should be made to assess the environmental consequences. When incinerators have capacities greater than 2,000 lb/h and include pollution prevention controls, an APC system should be installed, but it need not meet the high particulate removal efficiencies required of incinerators of the same capacity without pollution prevention controls.

Final determination of whether a medical waste incinerator with pollution prevention control should require APC equipment should be based on the results of an air quality modeling and health risk assessment. This assessment can be done individually for a specific system or performed in a generic manner with the results being extrapolated for similar systems. State regulatory agencies or the U.S. EPA could take the initiative in demonstrating the benefits of pollution prevention and in developing emission requirements for different size incinerators with pollution prevention controls.

6.4 AIR QUALITY MODELING AND HEALTH RISK ASSESSMENTS

Many states require that an air quality modeling and health risk assessment be conducted as part of the permitting process for medical waste incinerators. Air quality modeling uses EPA-approved computer models to estimate how the stack emissions, or stack plume, disperses from the incinerator stack. The air quality modeling analysis calculates how the plume from the incinerator stack affects ground level pollutant concentrations. Input into the model consists of meteorological conditions that simulate a wide range of weather conditions and provide a conservative estimate of worst-case pollutant dispersion. Other inputs into the computer air quality model include stack parameters (velocity, temperature, height, etc.), pollutant emission rates, local terrain features that may affect dispersion (tall buildings, hills), and a variety of parameters associated with atmospheric dispersion calculations, such as temperature gradients and wind velocity profiles.

Estimated pollutant concentrations from the air quality modeling analysis are compared to state and federal standards for ambient air quality concentrations to show compliance. In states where there are no established standards for pollutants, or where standards are only guidelines, a health risk analysis may be required to show that estimated pollutant concentrations will not significantly increase the risk to public health.

A health risk analysis for a medical waste incinerator estimates the incremental cancer risk due to the presence of the incinerator. The incremental cancer risk is the estimated excess probability of contracting cancer as the result of constant exposure over a 70-year lifetime to the ambient concentrations resulting from operation of the facility. An incremental cancer risk of less than one in a million is considered an acceptable risk. Maximum concentrations estimated in the air quality modeling analysis for specific pollutants of concern are multiplied by their unit risk factor to give the estimated incremental cancer risk.

Unit risk factors are based on carcinogenic potency factors established by scientists in the U.S. EPA's Cancer Assessment Group (CAG). The CAG recognizes that there are numerous uncertainties in the process by which it has determined cancer potency factors. The CAG's policy has been, when faced with uncertainty, to err on the side of overestimating the carcinogenic potency and, therefore, the carcinogenic risk. Thus, the CAG's policy is a conservative approach. A health risk analysis is based on a conservative methodology. Risks are based upon maximum concentrations estimated in the air quality modeling analysis using conservative modeling assumptions. Also, people living near an incinerator facility are not usually exposed to the concentrations assumed in the health risk analysis, because the incinerator will generally not operate for a 70-year period, and people are likely to spend significant time elsewhere. The health risk assessment can be an important analysis in showing the public the comparative risks associated with a specific medical waste incinerator. For medical waste incinerators with pollution prevention controls, the results would give important information concerning the requirements for APC equipment for different size systems. An important factor affecting pollutant concentrations downwind of a medical waste incinerator is the height of the incinerator stack. Higher stacks provide better dispersion, which lowers concentrations of pollutants downwind of the stack. This is taken into account in an air quality modeling analysis. Controlling pollutant concentrations by using a higher stack is not generally recommended due to the high costs of the stack, and because it is better to minimize pollutant emissions rather than disperse them (i.e., "dilution is not the solution to pollution"). Nevertheless, it is mentioned here because it is a method used to reduce downwind pollutant concentrations.

6.5 SUMMARY

Pollution prevention for medical waste incinerators can result in lower emissions and hence preclude the need for expensive APC equipment that many states are requiring for new and existing medical waste incinerators. In some cases, APC equipment may still be required for medical waste incinerators with pollution prevention controls, but the requirements should be less severe, thereby resulting in lower capital and operating costs for the APC equipment. Pollution prevention can lower pollutant emissions to safe levels without the need for APC equipment, and because there is no APC equipment, there is no disposal problem of potentially toxic fly ash or scrubber effluent.

The high cost of APC equipment makes it difficult for many hospitals to afford medical waste incineration, thereby forcing them to dispose of their waste via commercial waste haulers, who usually incinerate it at a high cost that is eventually passed on to the public. Pollution prevention, in addition to benefiting the environment, could also help hospitals financially. An air quality modeling and health risk assessment for different size medical waste incinerators practicing pollution prevention should be conducted to quantify the exact pollutant emissions and health risks. These health risks can then be compared to those for medical waste incinerators not practicing pollution prevention.

It should be noted that a large volume of APC design literature has developed (Work and Warner 1978; Brunner 1985; Cooper and Alley 1986; Brady 1989). The major current trend is to adapt these APC systems to medical waste incinerators (U.S. EPA 1985; Doucet 1986; Konheim and Hasselriis 1988; U.S. EPA 1990; Barton et al. 1990; Corbus 1990). A rigorous pollution prevention program at the hospital level should greatly lower the cost and reduce the complexity of the APC system needed to comply with state regulations.

REFERENCES

Barton, R. G., G. R. Hassel, W. S. Lanier, and W. R. Seeker. "State-of-the-art assessment of medical waste thermal treatment." EPA 68-03-3365, U.S. EPA, Research Triangle Park, NC, 1990.

Brady, J. D. "Economically and operationally attractive incinerator emission controls." Anderson 2000 Technical Report, Peachtree, GA, 1989.

Brunner, C. R. 1985. *Hazardous air emissions from incineration.* New York: Chapman and Hall.

Cooper, C. D. and F. C. Alley. 1986. *Air Pollution Control: A Design Approach.* Boston, MA, PWS Publishers.

Corbus, D. "A comparison of air pollution control equipment for hospital waste incinerators."Presented at the proceedings of the 83rd Annual Meeting of the Air and Waste Management Association, Pittsburgh, PA, June 1990.

Doucet, L. G. "Controlled air incineration: Design procurement and operational considerations." Prepared for the American Society of Hospital Engineering, Chicago IL, Technical Document No. 55872, January 1986.

Konheim, C. S. and F. Hasselriis. "Characterization of Hospital Waste Emissions." Presented at the proceedings of the 81st Annual Meeting of the Air and Waste Management Association, Dallas, Texas, 1988.

U.S. EPA 450/3-88-017. "Hospital waste combustion study." U.S. Environmental Protection Agency, 1985.

U.S. EPA 625/6-89/024. "Operation and maintenance of hospital medical waste incinerators (handbooks)." Cincinatti, OH: U.S. Environmental Protection Agency, 1990.

Work, K. and C. F. Warner. 1978. *Air pollution: Its origins and control.* New York: Harper & Row.

7

Ash Disposal

Floyd Hasselriis

7.1 INTRODUCTION

The history of ash disposal is significant to users and operators of medical waste incinerators (MWIs) in that they must deal with recently developed public perceptions and state regulations. Ash residues from MWIs are normally sent to sanitary landfills, which are permitted for the disposal of municipal wastes. Ash is usually placed in separate containers or in containers also used for general trash. The contents of these containers are picked up by contracted waste haulers, who transport the waste to landfills.

Concern about the nature of ash residues from municipal solid waste incinerators (MSWIs) has led to regulation of disposal of these residues by many states. In some states these regulations require testing the ash residues initially when the facility starts up and periodically thereafter. In some cases, the ash residues are required to be tested and the results reported before the residues may be removed from the site.

While the regulations have been developed for MSWIs having capacities of 100 to 3,000 t/d, they may be imposed on MWIs having capacities of 150 to 2,000 lb/h. MSWIs are usually operated 24 hours per day, 7 days per week, whereas MWIs are commonly charged for 6 to 12 hours per day and for 3 to 6 days per week. Despite this enormous difference in capacity, MWI operators often face requirements that do not take this into account.

The concern about the potential toxicity of MSW ash residues developed as the result of a number of historic reasons, which have exacerbated the concern. Opponents of incinerators cite data generally believed to show that ash residues are highly toxic.

Opposition to incineration has been used by many environmentalists as

142

a means to encourage recycling and by local residents who "just say no." These campaigns have been effective, forcing the states to require recycling goals to be set and carried out. Today solid waste plans are required by law, and ambitious goals ranging from 25 to 60 percent recycling have been set.

The campaigns against incineration originally focused on toxic air emissions. When vendors came through with highly efficient emission controls, reducing emissions by more than 99.9 percent, the emphasis shifted to the ash residues. Ash leaching data that were inconsistently and unscientifically collected were published, with the claim that the ash was toxic because many samples of bottom ash and mixed bottom and fly ash residues failed to pass the extraction procedure (EP) toxicity test, and that all fly ash samples failed. Since the EP test was designed to be a measure of toxicity, ash residues that failed the test were henceforth called toxic, and ash in general was often referred to as *toxic ash*. In addition, wastes that failed the EP test might be classified as hazardous wastes, requiring disposal at high cost in special double-lined landfills. Many proposed facilities were halted and some operating plants shut down after local opposition was aroused on the issue of toxic ash. Despite these unusual situations, however, most states classify MSW ash as a *special* waste, usually requiring disposal in separate landfills called *ashfills*. The U.S. EPA Guidelines and many states require only single liners to be used for ash residue monofills, as distinguished from the double liners required for MSW itself. Leachate collection is required, as well as proper treatment and disposal.

7.2 RESEARCH ON ASH LEACHING

Early studies of leaching from MSW landfills and incinerator ash residues produced variable and inconclusive data, mainly because the landfills investigated were existing facilities that did not have a means of collecting and hence of testing the leachate. This situation led to reliance on a testing procedure that was developed in a research laboratory. It was not known whether or not this procedure could predict the leachates actually produced in the field (U.S. EPA 1987, 1989).

The resource recovery industry, devoted to building and operating waste-to-energy facilities, when faced with severe problems in disposing of the ash, sponsored a major research program intended to settle, once and for all, the issue of whether ash residues are toxic and whether the U.S. EPA toxicity test predicted actual leaching conditions, and to obtain data that could be used to assure the public that methods of disposal would not contaminate drinking water and the environment with toxic metals.

The Coalition on Resource Recovery and the Environment (CORRE) sponsored a program to sample and analyze ash residues from five state-of-the-art waste-to-energy facilities, all of which have acid gas emission controls and use lime as the reagent (NUS 1990). The ash samples were collected in a uniform manner at all facilities. Several leaching procedures were employed in the analysis of the ash samples. The leachate from landfills accepting these ash residues was also tested, so that it could be determined which of several leach-testing procedures most accurately predicted the actual leachate composition.

The conclusions of the CORRE test program can be summarized as follows (NUS 1990):

- "None of the leachate samples exceeded the EP Toxicity Maximum Allowable Limits (MALs) established for the eight metals" classified as critical in drinking water standards.
- The data show that "although the leachates are not used for drinking purposes, they are close to being acceptable for drinking water use, as far as the metals are concerned."
- The ashes showed pH values of 10.36 to 11.85.
- The field leachate pH values ranged from 5.2 to 7.4.
- The ashes are rich in chlorides and sulfates. The total soluble solids in the ashes varied from 6,440 to 53,200 ppm (0.64 to 5.32 percent).
- The ashes contained unburnt total organic carbon (TOC) ranging from 4,060 to 53,200 ppm (0.4 to 5.32 percent).
- The TOC of the field leachate ranged from 10.6 to 420 ppm.
- Dioxins and furans (PCDDs/PCDFs) were found at extremely low levels in each ash sample, well below the Centers for Disease Control (CDC) recommended TCDD Toxicity Equivalency limit of 1 ppb in residential soil.
- The field leachate data showed that PCDD/PCDFs "do not readily leach out of the ash in the ash disposal facilities." They were found in the leachate from only one facility, probably originating from solids that were not properly filtered prior to analysis.

A number of extraction procedures were used, for comparison with each other and with field leachates, as follows (U.S. EPA 1989):

- Extraction Procedure Toxicity (EP toxicity test)
- Toxic Characteristic Leaching Procedure (TCLP 1 and TCLP 2)
- Deionized water (SW-924)
- Carbon dioxide procedure
- Simulated Acid Rain (SAR)

None of the extracts from the deionized water (SW-924), the CO_2, and the SAR extraction procedures exceeded the EP Toxicity Maximum Allowable Limits (MAL). The majority of the extracts from these three extraction procedures also met the Primary and Secondary Drinking Water Standards for metals. These procedures simulated the concentrations for lead and cadmium in the field leachates better than the extracts from the other three extraction procedures. (U.S. EPA 1988, 1989).

In summary, and contrary to previous beliefs, (1) the EP toxicity procedure did *not* simulate the leachates found at the landfills; (2) metals concentrations in the leachates were *far* below the U.S. EPA MAL, and, in fact, (3) the majority also met the standards for drinking water.

Despite the entirely new view resulting from this research, indicating that properly sampled and analyzed ash residues show that leaching of the critical metals should not be considered a problem, testing of ash residues will continue to be necessary in order to avoid having to characterize these residues as hazardous wastes, requiring disposal in special, more costly disposal facilities.

7.3 METALS CONTENT OF WASTES

Most natural materials, such as wood and minerals, contain at least trace quantities of metals. They are transferred to paper and are present in the fillers used in plastics. Figure 7–1 shows a typical total chemical

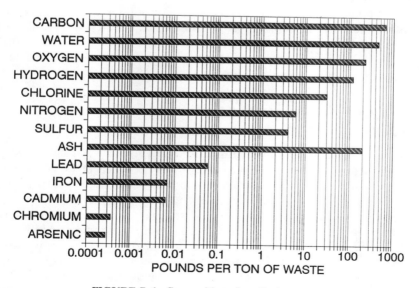

FIGURE 7–1. Composition of medical waste.

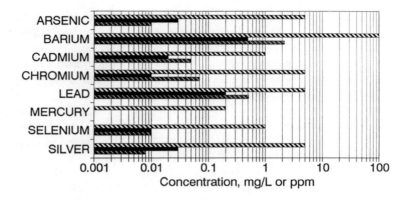

FIGURE 7–2. Metals leached from municipal and medical waste ash by the TCLP test.

composition of municipal or medical waste, which, roughly speaking, is made of similar materials. In the ash, shown as 200 lb/t, or 10 percent of the total, the trace metals include lead, at 0.056 lb/t, and cadmium, at 0.0065 lb/t, down to arsenic, at 0.0003 lb/t. The metals content of medical waste is presented in greater detail in Chapter 2.

The greater part of these metals remains insoluble after combustion, whether they end up in the fly ash or the bottom ash. However, about 5 percent of the lead in the fly ash may be in soluble form, and more than 50 percent of the cadmium. The solubility of these toxic metals is the main threat to the environment and health.

7.4 LEACHING OF METALS IN ASH RESIDUES

The degree of leaching of the metals can be measured by various methods, as discussed above. Although the improved TCLP test is no longer considered to be valid for predicting environmental behavior, it will probably continue to be used for some time.

The results of TCLP leaching tests of ash residues produced by incinerators burning municipal waste and medical wastes are compared in Figure 7–2, and compared with the U.S. EPA MAL. As an example, for lead, the MWI data showed 0.5 mg/l (0.5 ppm by weight), whereas the MSW ash showed 0.2 ppm. These were both well below the allowable MAL of 5.0 ppm. Note that the MAL is 100 times the standard for drinking water.

Although the data in this graph show that the MWI ash exhibited higher leaching of metals, this cannot be assumed to be true in general. The MSW incinerator from which this ash was obtained is a starved-air incinerator, similar to most medical waste incinerators. Due to the lower furnace temperatures used, starved-air incinerators can be expected to have much lower emissions than excess-air, waterwall combustors, since less of the volatile metals may be emitted as fly ash.

7.5 EFFECT OF FURNACE TEMPERATURE ON PARTICLE SIZE

The particle size distribution of fly ash from MWIs is of concern for several reasons. First, the collection efficiency of emission control devices depends upon the particle size. The emitted particles are likely to be the smaller ones that fly with the air and do not settle rapidly, if at all. Next, the fine particles are not removed by the collectors of the human respiratory system, hence they can get through into the lungs. Since the acids and metals on the particles are readily absorbed by the lungs, these fine particles are of the greatest concern. Particles smaller than 1 μm are considered to be in the respirable range. The particle size distribution depends strongly on the temperatures used in the primary and secondary combustion chambers, since it is here that the metals are vaporized.

Recent research has shown clearly that high furnace temperatures greatly increase the quantity of volatile metals carried from the combustion chambers with the fly ash. Data from these studies are presented in Chapter 5. The size distributions of the particulate emissions were found to vary substantially when the secondary furnace outlet was maintained at different (average) temperatures (see Figure 7–3). The fly ash that was smaller than 1 μm varied from 15 percent at 1,350°F to 65 percent at a high temperature range averaging 2,150°F. Analysis of the fly ash showed that this difference was clearly due to metal aerosol, which makes extremely fine particulate (Brady 1991). State requirements of temperatures of 1,800°F or higher would thus appear to cause a high degree of volatilization and emissions of toxic metals such as cadmium.

7.6 VARIABILITY OF ASH LEACHING DATA

One of the main reasons that informed persons were concerned about the toxicity of ash residues was the obvious high variability of analytical test results. As compared with the MAL of 5 ppm for lead and 1 ppm

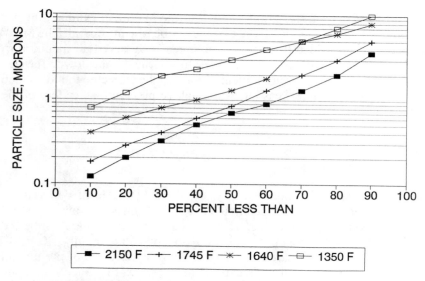

FIGURE 7–3. Particle size distribution versus temperature.

for cadmium, fly ash samples often exhibited more than 10 ppm for lead and 2 ppm for cadmium. These "hot spots" were cited as evidence of toxicity, even when the average of a number of samples was below the MAL.

To address this high variability, it is necessary to understand more about statistics and the science of obtaining representative samples. One sample could give a reading of anything from 0.01 to 30 ppm of lead, for instance. Until five or more samples are analyzed, there will be no basis for estimating the true average or mean value. Therefore, only multiple samples and multiple analyses will provide confidence that the mean value has been approached.

7.7 RESEARCH ON MSW FLY ASH AND BOTTOM ASH RESIDUES

Early research on ash samples from municipal incinerators showed that the fly ash contained high levels of lead, especially soluble lead, as well as cadmium, whereas the levels in bottom ash were very low. The reason for this is that the metals vaporized during combustion of a waste containing chlorine are likely to form chlorides in the fly ash, which are highly soluble. Since the fly ash had higher soluble lead content, it would be important to thoroughly mix fly ash with bottom ash to prevent hot spots from showing up, especially in laboratory samples.

As a result of this finding, procedures have been established for obtaining well-mixed, representative samples of ash. They are as follows:

- Collect full-shovel samples from across the belt.
- Take samples hourly for at least one shift.
- Cone and quarter the day's sample down to the laboratory and duplicate samples (about 1 kg each).
- Screen the sample through a 2-inch screen to remove oversize objects. Screen the remainder through a 3/8-inch screen. Put the oversize through a shredder, and screen again. Weigh the rejected unscreenable material as a fraction of the whole sample, to apply a correction later.
- The laboratory reduces the sample further, with additional grinding. When doing the leaching test, be careful not to add the acid too fast, since this would dissolve too much metal.

The potential toxicity of fly ash derives from the solubility of the metals, not from the insoluble chemical forms. When fly ash falls on the ground, the soluble metals will leach out readily. When inhaled, they are readily absorbed by the body. When fly ash is wetted, it agglomerates readily and will not return to fly ash, in the same manner that lime dust when wet makes a good paint, which is hard to dust off. When fly ash is mixed with bottom ash, it not only paints the larger particles, it forms a cement that encapsulates the heavy metals. For these reasons, the best way to dispose of fly ash is to get it wet as soon as possible and mix it with damp bottom ash. This is the procedure generally followed with residues from both municipal and medical waste incinerators.

7.8 ASH RESIDUES FROM MWIs

Ash samples from periodically analyzed MWIs show the same variability as those from MSWIs. This statement is supported by the results of analysis of samples collected over a 6-month period at a hospital in Florida.

Figure 7–4 shows the data presented as statistical distribution. This means that the data are sorted into an ascending series and spaced apart at equal distances on a logarithmic scale. The fact that nearly straight lines result indicates that the distribution is log-normal. The consistency of the data is shown by how well the points fit the straight lines.

The samples of water taken from the ash quench tank show a remarkable consistency. Reading the graph, we see that only 13 percent (one of every eight) of samples would contain more than 1 ppm of lead, and the average

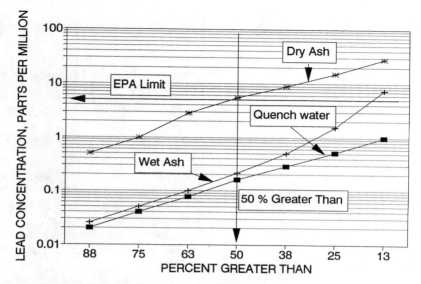

FIGURE 7–4. Soluble lead in medical waste ash and quench water.

would contain about 0.15 ppm of lead. This is well below the EPA limit of 5 ppm for lead.

The wet ash, drawn out of the quench tank, was leached by the EP toxicity procedure. This procedure uses 20 parts of water to leach one part of ash (all by weight) and tumbles it overnight. Acid is added to the water to maintain a pH of 5.0. This is the *acid test,* which, as noted above, is more extreme than what happens in a landfill. Note that one sample at 7.5 ppm somewhat exceeded the MAL of 5 ppm, but 50 percent of the samples showed less than 0.2 ppm.

The leachate from dry ash residues represents ash that was removed from the incinerator *after* the overnight burndown and not passed through the quench tank. It thus contained no carbon and no moisture. The lead concentrations from this ash ranged as high as 28 ppm, averaging close to the MAL of 5 ppm.

Assuming that the normal wet ash contains glass, metal, moisture, and carbon totaling 65 percent, this dry ash is only 35 percent of the wet ash. Naturally, it is more concentrated. Correcting for this concentration, the dry ash would have shown a maximum of 10 ppm, and the average would be 2 ppm. While these are higher concentrations than the wet ash, the average would still be far less than the MAL. After concentration is corrected, it is still evident that the fly ash contains more acid-soluble metal than the wet ash, hence must be given special consideration.

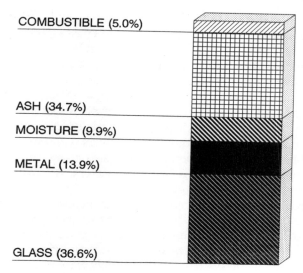

COMBUSTIBLE (5.0%)

ASH (34.7%)

MOISTURE (9.9%)

METAL (13.9%)

GLASS (36.6%)

FIGURE 7–5. Total composition of ash residues from a medical waste incinerator.

7.9 COMPOSITION OF MWI ASH RESIDUES

A complete analysis of the ash residues shows the many components that must be taken into account. The sampling and analysis procedures must correct for the glass, metals, and slag that are not correctly included in the ash analysis. A correct procedure screens out the oversize insoluble components of a large ash sample and calculates their percentage back into the analytical results after the screened residues are analyzed, using the procedure outlined above.

Figure 7–5 shows the complete analysis of an incinerator ash sample collected from a medical waste incinerator in New York City. The moisture of the entire sample was only 10 percent. However, after deducting the 50 percent glass and metal, as shown in Figure 7–6, the moisture fraction increases to 20 percent. The carbon in the ash was 5 percent in this case.

Typical ash residues drawn from a quench tank usually contain 40 to 70 percent dry solids and 30 to 60 percent water. The variation reflects the amount of carbon in the ash, since it holds its weight in water.

Incinerators having quench tanks provide the opportunity to wash the ash prior to disposal. It has been found that the hot water in the quench tank is effective in dissolving heavy metals in the ash residues. The ash residues removed from the tank after washing may be almost entirely free of soluble metals. The main drawback is the need to supply fresh water and to treat or dispose of the salty water. With washing, the carbon and other unburned materials can be floated off and reburned (Hasselriis 1991).

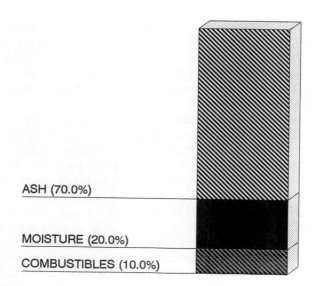

FIGURE 7–6. Composition of ash residues from a medical waste incinerator with metals and glass removed.

7.10 EFFECT OF EMISSION CONTROLS ON ASH DISPOSAL

Two types of emission controls are used to remove particulate matter and acid gases: wet scrubbers and dry lime scrubbers with baghouses. Wet scrubbers absorb the acids to form sodium chloride, which, together with the soluble metal chlorides, is generally discharged to the sewer. For small systems this discharge of salty water can usually be permitted since it is diluted with such a large amount of other water wastes before it leaves the premises. Systems are available for eliminating wet discharges and reducing the small mass of salts to dry form for disposal in a hazardous waste landfill.

Baghouse emission controls produce a dry powder that has a high calcium hydroxide (lime) content. This has a high alkaline excess, so the pH may be higher than 11. At these high pH values the lead can become soluble. If the fly ash is mixed with bottom ash the alkalinity will be reduced, preventing the lead from becoming as readily soluble.

7.11 REDUCING SOLUBLE METALS AND SALTS IN ASH RESIDUES

From the above discussion, several ways of reducing the environmental risk of ash residues can be noted:

1. Reduce the lead and cadmium in the waste to reduce the amount of soluble lead and cadmium in the fly ash.
2. Reduce the amount of chlorine in the waste to reduce the amount of soluble chlorides in the fly ash.
3. Do not operate the combustion chambers at very high temperatures. Since a range from 1,600°F to 1,800°F has been found to be optimum for destruction of organics and production of the least carbon monoxide, the lower end of this range (<1,800°F) would be best to minimize heavy metals volatilization.

By these methods, the amount of soluble and, hence, toxic forms of these metals could be reduced in the fly ash and therefore in the mixed fly ash and bottom ash.

7.12 SUMMARY

Ash residues contain the heavy metals that were originally in the waste materials. Only a small portion of the metals in the bottom ash is likely to be soluble. However, a much larger percentage of the metals in the fly ash may be soluble because they are formed as salts of chlorine. Since metals are vaporized and discharged with the fly ash at high furnace temperatures, incinerators should not be operated at excessively high temperatures.

Dry fly ash is a toxic material and should be conditioned with water or moist ash as soon as possible in order to prevent harm to humans and the environment. When the fly ash is mixed with damp bottom ash, in addition to the dilution that takes place, chemical reactions occur that tend to make the potentially toxic ash much less soluble. For this reason, when the ash residues are landfilled, very little of the metals is found to leach out into the collected leachate.

It is important to collect ash samples in such a manner that the samples are representative of the ash disposed of in the landfill. Tests of ash residues from medical waste have shown that the leachates from the TCLP test meet the U.S. EPA limits, on average, although the data may vary widely. The water in the quench tank was also found to meet the U.S. EPA limits in the instance cited. The dry ash residues remaining after burndown were found to have the highest leachable lead content.

Even though the ash residues become relatively benign after wet reactions have taken place, the salts that result from the neutralization of hydrogen chloride may create additional burdens on the processes needed to treat the leachate before it can be returned to water supplies. Thus ash residue problems as well as toxic emission problems can be mitigated by reducing the amount of chlorinated materials in the wastes slated for incineration.

REFERENCES

Brady, J. D. "Recent developments in pollution control systems for chemical and infectious waste incinerators." Presented at Mid—West American Institute of Chemical Engineers Meeting, St. Louis, MO, February, 1991.

ELI. "Report of results—waste incinerator ash sample, New York City Hospital" Environmental Laboratories, Inc., Glen Cove, NY, 1985.

Hasselriis, F. "Variability of municipal solid waste emissions from its combustion." American Society of Mechanical Engineers Solid Waste Processing Division Conference, Orlando, FL, 1984.

Hasselriis, F. "Relationship between municipal waste combustion conditions and trace organic emissions." Presented at the 78th Annual Meeting of the Air Pollution Control Association, Detroit, MI, June 1985.

Hasselriis, F. "Minimizing refuse combustion emissions by combustion control, alkaline reagents, condensation, and particulate removal." Synergy/Power Symposium, Washington, DC, October 1986.

Hasselriis, F. 1987. "Optimization of combustion conditions to minimize dioxin emissions." *Waste Manage. & Res.,* 5(3): 311–26.

Hasselriis, F. "Variability of municipal solid waste and emissions from its combustion," American Society of Mechanical Engineers Solid Waste Processing Division Conference, Orlando, FL, 1984.

Hasselriis, F. "Relationship between municipal refuse combustion conditions and trace organic emissions," Presented at the 78th Annual Meeting of the American Pollution Control Association, Detroit, June 1985.

Hasselriis, F., et al. "The removal of metals by washing of incinerator ash." Presented at the 84th Annual Meeting of the Air and Waste Management Association, Vancouver, BC, June 1991.

NUS. EPA/530–SW–90–029A, NTIS PB90-187154 "Characterization of municipal waste combustion ash, ash extracts, and leachates—Coalition on Resource Recovery and the Environment." U.S. EPA Office of Solid Waste, Washington, DC, 1990.

U.S. EPA. EPA 530–SW–87–028 "Characterization of MWC ashes and leachates from landfills, monofills, and co-disposal sites." U.S. EPA Office of Solid Waste, Washington, DC, 1987.

U.S. EPA. "National secondary drinking water regulations." *Federal Register 40-CFR 143,* September 1988.

U.S. EPA "National primary drinking water regulations." *Federal Register 40-CFR 141,* April 1989.

8

Polymer Substitutes for Medical Grade Polyvinyl Chloride (PVC)

Kenneth B. Wagener, Christopher D. Batich, and Alex E. S. Green

8.1 PROPERTIES OF PVC

Commercial production of PVC began in 1933 in both Germany and the United States. Although the polymer had been known since 1872, it had never found use because of severe problems with decomposition during processing. Because of the combination of PVC with plasticizers and other property modifiers, it has become the most widely used polymer in the world now in terms of the variety of applications. The material is made from vinyl chloride monomer (Figure 8–1), which is a gas at room temperature. This monomer can be copolymerized with a variety of other monomers yielding a host of other polymers containing significantly reduced amounts of vinyl chloride (*Modern Plastics Encyclopedia* 1990; Saunders 1988).

PVC has been the most commonly used polymer in the medical device arena (Anonymous 1984). Hence, a large percentage of it is included in the incineration waste stream. It has been estimated that medical grade PVC will grow 6.4 percent per year to about 700 million pounds in 1994, and the main emphasis will be on the commercialization of new medical grades and more easily sterilizable grades. Among the main uses are IV drug delivery, catheterization, and blood administration (Anonymous 1991).

The properties of PVC are excellent as far as functionable use is concerned. Once formulated correctly, it can be extremely tough and transparent and can act as a good barrier to diffusion (Hinkling 1991). The glass transition temperature can be shifted to an appropriate range by adding plasticizer. These properties also make PVC an outstanding coating material, and it has found use in this area also. Hence, the replacement of PVC

155

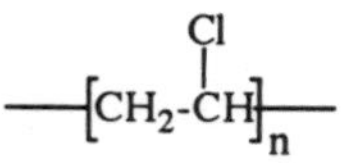

$$-[CH_2-\underset{\underset{Cl}{|}}{CH}]_n-$$

FIGURE 8-1. Polyvinyl chloride (PVC).

will require a variety of polymers to replace the variety of functions PVC now assumes. Such replacement materials are available.

PVC is widely used in the medical industry for three pragmatic reasons. First, the polymer has been readily available for many years, so a great deal (positive and negative) is known about it, and the techniques necessary to shape the polymer also are widely known. Second, the polymer is quite inexpensive, both in its raw state and in its blended state with plasticizers, and as a consequence, its utility in the marketplace has been investigated for many applications, not only medical ones. Third, the physical properties of plasticized PVC are very inviting and generally fit the demands of medical applications very well. These properties can be varied depending upon the weight percentage of plasticizer that is added to the polymer to achieve the required degree of flexibility.

Polyvinyl chloride is normally synthesized in its atactic form, and as a consequence, the polymer is mostly noncrystalline and brittle in its native state. This brittleness is attributed directly to the fact that the glass transition temperature of PVC is above its normal use temperature. While this fact has been long known, it has also been known that plasticizers may be added to PVC in order to overcome this brittleness property, and in fact, if sufficient plasticizer is added to native PVC, the product is a very flexible material that can be molded and shaped as desired. The range of plasticizers investigated over the years is quite wide, and for reasons of availability and cost, phthalate plasticizers are the preferred materials. For example, dioctylphthalate (DOP) is widely used as a plasticizer in PVC and is added as much as 30 wt% or more to the native polymer in order to achieve a material with flexibility suitable for medical applications.

The nature of the plasticizer/polymer interaction is not completely understood, yet it is clear that a homogeneous blend of plasticizer and polymer results. This homogeneous blend leads to a material that has long-term durability, stability to light and heat, and normal atmospheric exposure. Thus, PVC is quite a useful material in the medical field and has proven its worth through its success and application over a long time.

On the other hand, the very plasticizer that is added to achieve flexibility also leads to an inherent problem associated with the use of PVC in the medical field. The plasticizer is not chemically bound to PVC, and thus it is susceptible to extraction by exposure of the polymer to liquids. Consequently, the plasticizer can be slowly extracted by either flowing liquids in tubes and the like or in storage bags containing liquids of medical interest. For this reason, finding substitute materials that are inherently flexible would be a useful venture to pursue. Inherently flexible materials do exist, yet often they are more expensive, and equally as often, the range of flexibility (or desired material properties) does not quite match that of plasticized PVC.

8.2 COMBUSTION PRODUCTS OF PVC

The major problems associated with PVC become most noticeable when the polymer is incinerated. For red-bag waste (potentially infectious medical waste), incineration is now the primary logically available option, hence it is important to consider substitute materials. Perhaps the most straightforward problem with PVC is the large amount of chlorine contained in the polymer. For pure PVC, 56.72 percent of the weight is chlorine, 38.44 wt% is carbon, and 4.84 wt% is hydrogen. When 1 lb of PVC is incinerated, 0.58 lb of hydrogen chloride is produced. The corrosion potential of this product might be visualized by noting, by way of comparison, that muriatic acid used for cleaning swimming pools is 30 percent hydrochloric acid and 70 percent water. This can be a huge burden on the disposal system. As noted in Chapter 6, scrubbers can remove the acid but then something must be done with the large quantities of calcium chloride formed. In addition, the scrubbers frequently cost as much as the incinerator itself. This is why most hospitals currently do not have acid scrubbers with their incinerators.

In addition to corrosion problems, toxic chlorinated hydrocarbon emissions are usually associated with the incineration of PVC. In general the potential toxic emissions generated by incineration of plastics depends upon the atomic composition of the polymer, the additives used in formulating this plastic, and the combustion conditions in the various incinerator stages. Boettner et al. (1973) attempted to identify toxic products of incomplete combustion (PICs) in accidental plastic fires and to identify information needed to minimize PICs in the incineration of plastics. They studied the combustion of polyvinyl chloride, polysulfone, polyurethane, polyimide, Lopac®, Barex®, phenol formaldehyde, urea formaldehyde, polyethylene, polypropylene, polystyrene, polycarbonate, polyphenylene oxide, polyester, synthetic fabrics (Dacron®, Orlon®, nylon), and natural products (wood and wool). In their experimental system 1- to 3-g samples were heated at a controlled rate from 5 to 50°C/min in the presence of a measured flow of air or air plus oxygen. By this method many gaseous and condensed PICs were produced. Their results, while more representative of emissions

from accidental fires, are nevertheless of interest in the design and operation of medical waste incinerators. Their chromatographic analysis of PVC
combustion products detected about 60 volatile products. Of these, 52
compounds have been identified including the major products carbon dioxide, hydrogen chloride, and carbon monoxide, which were present at 0.7
to 0.1 g/g. Next came benzene, acetic acid, methane, ethane, propane,
and vinylchloride, which typically were found at 40 to 1 mg/g levels. Their
study also showed 40 or so complex hydrocarbons at levels of 0.1 to
0.01 mg/g. Analysis for benzyl chloride was conducted on five commercial
meat-wrap films under various heating procedures giving results ranging
from 0.1 to 5 mg/g.

Recently Pasek and Chang (1991) specifically examined the thermal
degradation of PVC and polyvinylidene chloride (PVDC) in an endeavor
to determine the role played by plastics in forming toxic chlorinated compounds in incinerators. Approximately 10 to 20 mg of PVC resin were
placed into a pellet press and compressed at about 20,000 lb/in^2. The pellet
was loaded into a Dupont 951 Thermogravimetric Analyzer (TGA), which
measures the weight loss of the pellet as it is heated at a programmed rate
in a controlled atmosphere. The TGA was purged with 99.998 percent
purity N_2 to provide a pyrolytic ambience. As the purge gases and decomposition product exit the TGA, a small slipstream is pulled across a Supelco
Carbotrap™ 200 dual bed sorbent tube at a rate of about 20 ml/min. The
products captured on the trap were analyzed by gas chromatography/
mass spectography. Their results indicate that, apart from HCl, very few
chlorinated compounds are directly formed in the case of PVC thermal
degradation. However, in the case of PVDC many chlorinated compounds
are generated directly. Figure 8–2 shows a mass selective detector (MSD)
trace of degradation products of PVC. Figure 8–3 shows the corresponding
trace for PVDC. Chlorinated hydrocarbon stack emissions from PVC
combustion nevertheless could result from the chlorination mechanisms
discussed in Chapter 1 acting upon products of incomplete combustion.
In the case of PVDC, chlorinated hydrocarbons can result directly from
incomplete combustion.

The research of Pasek and Chang (1991), Boettner et al. (1973), and
others indicate initial mechanisms for the production of chlorinated and
nonchlorinated aromatic hydrocarbons that can serve as precursors of
PCDDs and PCDFs. It is very important to note that the chemical transfor-
mations that actually take place in a full-scale incinerator or in uncontrolled
flames (Hilado 1990) can be very different from these in laboratory combustion or pyrolysis arrangements. Thus, one must interpret such results in
the context of engineering experience with real incinerators such as have
been described in Chapters 1 to 5.

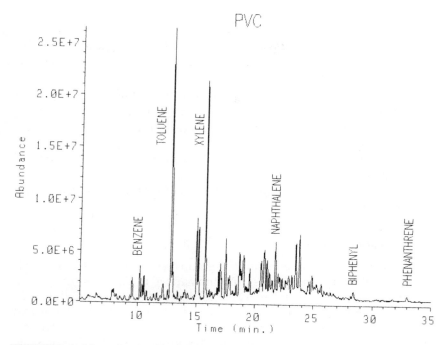

FIGURE 8–2. Mass selective detector trace of PVC thermal degradation. (From Pasek and Chang 1991.)

In addition to serving as potential sources of toxic chlorinated organic compounds, PVC plastics frequently contain lead- and cadmium-based stabilizers that can produce toxic metal emissions during burning. The chlorides produced by burning also react with other compounds such as lead and especially the chromium found in 316L stainless steel frequently used for hypodermic syringe needles. In this reaction, a metal chloride produced has greater volatility than the normal oxides present. Hence, more of these metals will be emitted. This is especially a concern for chromium.

8.3 SUBSTITUTE POLYMERS FOR FLEXIBLE MEDICAL APPLICATIONS

One longstanding application for plasticized polyvinylchloride is in the construction of blood bags because of the low cost of the material, its acceptable biocompatibility, and the excellent mechanical behavior of plasticized PVC. It can withstand high stress, and it also can be manipulated easily through heat-sealing techniques to construct blood bags of

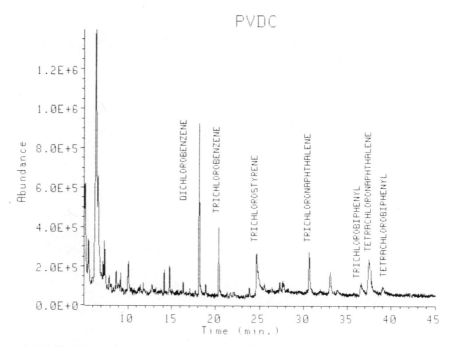

FIGURE 8–3. Mass selective detector trace of PVDC. (From Pasek and Chang 1991.)

various sizes. It is also known that plasticized PVC is compatible with other polymers that may be used in the construction of the bag itself such that heat sealing across materials is possible.

While PVC is indeed useful and the emission problems mentioned earlier remain under scientific debate, it is obvious that an inherently flexible material (one not requiring a plasticizer) would be preferred over plasticized PVC, if it were commercially available, relatively inexpensive, and able to meet the rigorous mechanical demands placed on such material for storing blood. In addition to these mechanical demands, the material must also be biocompatible and must not alter the composition of blood.

It has been known for some time that biocompatible materials for contact with blood or tissues can be made from block copolymers, specifically copolymers alternating *hard* and *soft* segments. The hard segments provide for mechanical strength needed in the design of a flexible material, while the soft segments confer inherent flexibility and biocompatibility. Since these so-called segmented copolymers exhibit inherent elasticity, the need for plasticizers is avoided. They also contain no chlorine. Thus, segmented or block copolymers would appear to be candidate materials

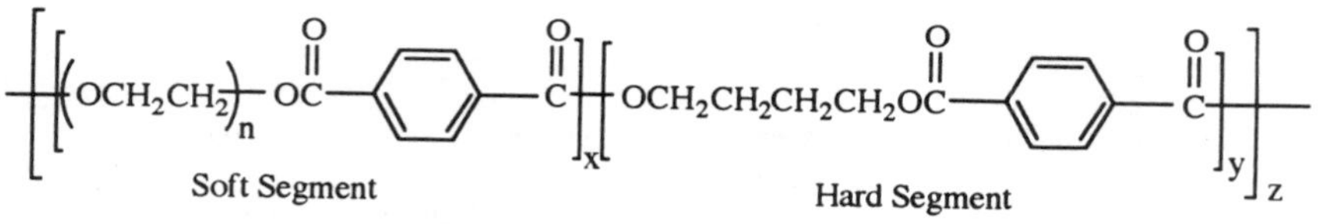

FIGURE 8–4. A polyester ether copolymer.

as a replacement for PVC in the construction of blood bags, and perhaps tubing as well.

Commercial segmented copolymer materials are available and are in fact known to be biocompatible, such as polyether polyurethanes (Biomer™) from Ethicon and polydialkylsiloxane polyurethanes (Avcothane™) from Avco Medical Products. Further, polyester/polyether block copolymers have been examined for their blood compatibility and have been shown to be bland with respect to their contact with blood. The term *bland* implies in this case that protein absorption and platelet absorption on the surface of the material is minimized.

The polyester ether copolymer shown in Figure 8–4 could be a candidate material for blood bags and other flexible PVC applications. These polymers are thermoplastic segmented polyester ethers that have long-chain ester units and short-chain ester units joined through ester linkages, and because of the nature of the short-chain ester unit, the polymers are crystalline, and consequently, quite strong. This crystalline nature of the polymer permits it to be shaped from the "melt" such that extrusion of film is quite possible. Thus, films can be made of polyester ethers that would be subject to heat sealing, thereby leading to the potential of constructing blood bags from these materials.

It is thought that the greater content of the soft segment on the surface of the polymer, the greater the blood compatibility of the copolymer itself. This observation suggests that the excellent phase separation, which occurs in polyester ethers, will lead to improved blood compatibility. Phase separation is a reasonably well-understood phenomenon, one that can be controlled by cooling rate, and it is likely that polyester ether materials will provide very bland surfaces with the contact of blood. Polyester ethers are indeed interesting candidates for the construction of blood bags.

Flexibility is an important property for any material in constructing blood bags, and it should be noted that the flexibility of polyester ethers can be varied by changing the amount of soft segment present within them. In a sense, the soft segment is a chemically bound plasticizer that allows for the flexibility observed. Thus, the more soft segment present, the softer the material, and the more flexible it becomes. This means that the physical

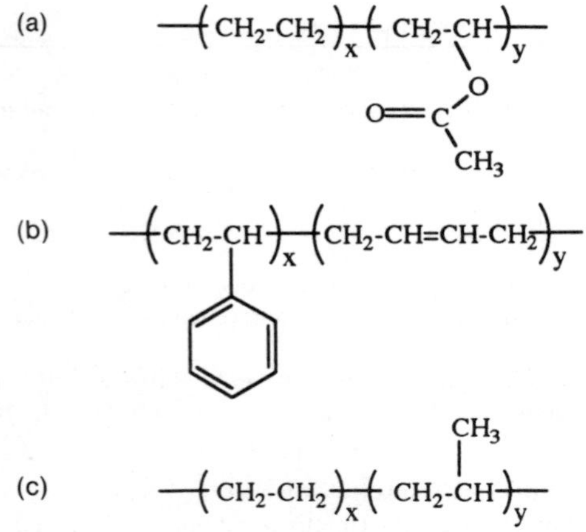

(a) $-\left(CH_2\text{-}CH_2\right)_x\left(CH_2\text{-}CH\right)_y-$

$$O=C\overset{O}{\underset{CH_3}{\diagup}}$$

(b) $-\left(CH_2\text{-}CH\right)_x\left(CH_2\text{-}CH=CH\text{-}CH_2\right)_y-$

(c) $-\left(CH_2\text{-}CH_2\right)_x\left(CH_2\text{-}\underset{\underset{CH_3}{|}}{\overset{CH_3}{|}}CH\right)_y-$

FIGURE 8–5. (a) Ethylene vinyl acetate elastomers; (b) polystyrene block elastomers; (c) ethylene propylene–based elastomers.

behavior of these polymers can be varied and perhaps tailored to the required flexibility that is needed in these materials.

It is important to note that polyester ether polymers are not at all esoteric, but in fact are commercially available and widely used in the engineering plastics industry. Polymers similar to the structure shown above are sold by DuPont in various grades of softnesses and could be considered as candidate materials for the construction of blood bag devices.

It should also be noted that these polyester ether materials consist of carbon, hydrogen, and oxygen, and consequently they represent environmentally sound materials for incineration. The incineration products are carbon dioxide and water when combustion is complete, and consequently they pose no greater risk to the environment than the incineration of natural polymers such as cellulose or other synthetic polymers such as polypropylene or polyethylene. This added advantage, in addition to being free of plasticizers, suggests that polyester ethers should be screened for utility in the construction of blood bag materials.

Polyester ethers do not represent the only polymers that might be considered in substituting PVC for blood bags, tubing, and other flexible uses. For example, the polyurethane ethers have been widely known in the medical industry and might have applicability here as well. The absence of chlorine in polyurethane ether structures is evident, and the advantages therein also are clear. Figure 8–5a shows ethylene vinyl acetate copolymers that have been available commercially for many years. Figure 8–5b illustrates polystyrene block elastomers on butadiene and other soft phases

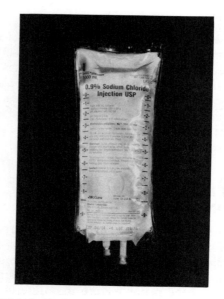

FIGURE 8–6. Flexible bag made from M-4680 olefin alloy.

that also are commercial. Finally, we note that ethylene propylene elastomers (Figure 8–5c) are important flexible materials and also have the potential for application in the medical industry. In fact a proprietary polyolefin (hydrocarbon) produced by Horizon Polymers (a Division of Ferro Corporation) has been used commercially in many flexible bag applications. Flexible bags made from this M-4680 olefin alloy are available from McGraw Co. for use with IV fluid administration (see Figure 8–6). Since each fluid must be compatibility tested for FDA approval, it has taken several years, but many fluids are already approved. This product shows that a flexible plastic that converts to carbon dioxide and water upon incineration is already being used by many hospitals in the United States.

8.4 SUBSTITUTE POLYMERS FOR RIGID PVC MEDICAL MATERIALS

Replacement of rigid PVCs is one of the easier substitutions to consider. A host of transparent, tough, easily processible thermoplastics are available that are not significantly more expensive than PVC or may in fact cost less. In 1984 there was already significant interest in finding substitutions for PVC in this market. Polypropylene is one of the most direct

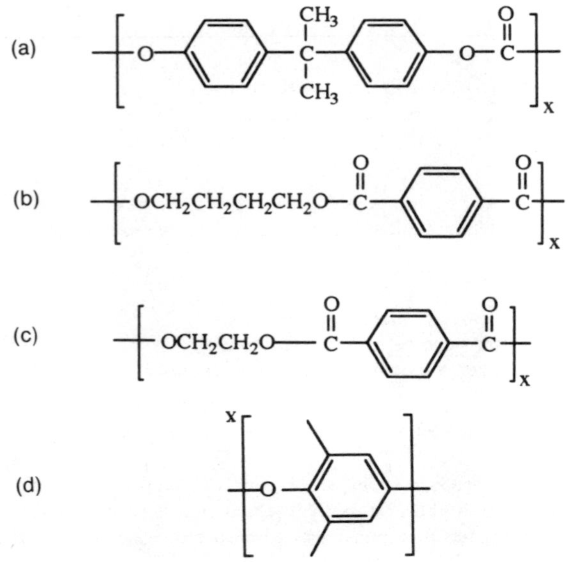

FIGURE 8–7. (a) Polycarbonate (PC); (b) polybutylene terephthalate (PBT); (c) polyethylene terephthalate (PET); (d) polyphenylene oxide (PPO).

replacements and is finding increased use in many areas. This is especially evident since polypropylene copolymers have been introduced by Arco mainly as a replacement for vinyl. In addition, vinyl elastomers may be replaced by polypropylene copolymers and hence will bond more readily to the polypropylene rigid polymers.

Some engineering (more expensive, but better properties) thermoplastic resins available include polycarbonate (PC), polybutylene terephthalate (PBT), polyethylene terephthalate (PET), and modified polyphenylene oxide (PPO). All of these polymers produce only carbon dioxide and water if incinerated properly. These structures are shown in Figure 8–7. In addition, acrylonitrile-butadiene-styrene (ABS), nylon-6, and polyphenylene sulfide are illustrated in Figure 8–8. Other potentially suitable rigid polymers include acetal, polysulfone, polyetherimide, polyetheretherketone (PEEK), and polyphenylene sulfide. It should be noted that the incineration of nylon and ABS produces nitric acid precursors, while polysulfone and polyphenylene sulfide produce sulfuric acid precursors.

Polystyrene resin (Figure 8–9a) currently available (e.g., from Dow) is a material with high clarity and processability. There are also copolymers of styrene; for instance, Phillips has a butadiene/styrene copolymer called K-resin that has significant property enhancements. These two materials

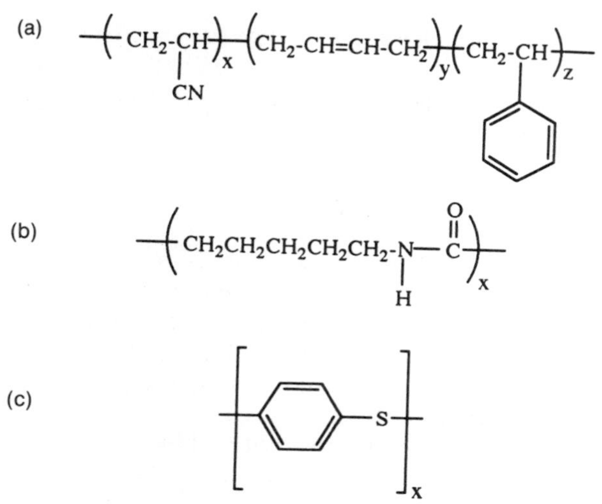

FIGURE 8–8. (a) Acrylonitrile-butadiene-styrene (ABS); (b) nylon-6; (c) polyphenylene sulfide.

are among the lowest cost materials available as substitutes for rigid PVC and are closely followed by grades of acrylic (Figure 8–9b), another low cost resin, ABS, SAN, and then the more costly polycarbonate and other speciality resins. Most issues of *Modern Plastics* carry articles on the current changes going on in plastic processing and substitution. The *Modern Plastics Encyclopedia* is an excellent source for basic information on plastics and where one may contact suppliers and distributors. Many more resources could be listed, and there are many manufacturers for most of the items discussed here. As noted in Section 8.2, the products of incomplete combustion of many plastics have been studied by Boettner et al. (1973). Careful attention must be given to emissions from incineration of plastics that might serve as substitutes for PVC. In particular, the plasticizers, additives, fillers, and pigments must be selected with care lest they generate toxic emissions as harmful as those from plasticized PVC.

8.5 ESTIMATES OF EXTRA COSTS

PVC is a current price performance leader. This is because of the huge economy of scale that exists for making PVC available at relatively low cost. PVC is the second largest mass production polymer in the world, immediately behind low density polyethylene. If there is a larger demand for these other resins, the economies of scale could significantly reduce the cost of production.

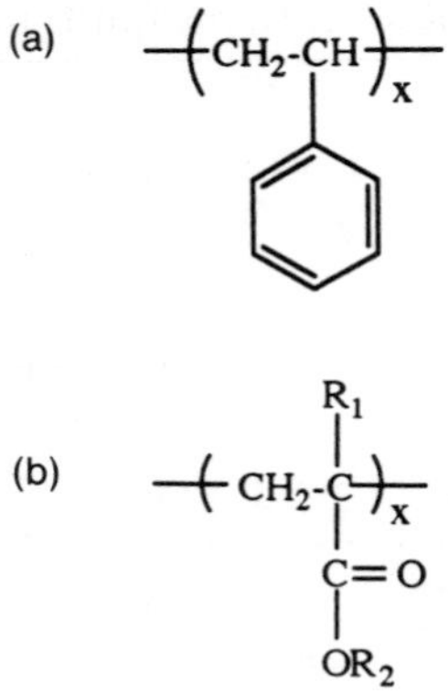

(a) $\left(\text{CH}_2\text{-CH}\right)_x$

(b) $\left(\text{CH}_2\text{-}\underset{\underset{\displaystyle OR_2}{\displaystyle |}}{\overset{\overset{\displaystyle R_1}{\displaystyle |}}{\underset{\displaystyle C=O}{C}}}\right)_x$

FIGURE 8–9. (a) Polystyrene; (b) acrylic polymers.

However, the cost of the resin involved in making a medical device is relatively small. In other words, the raw materials cost for most of the plastics involved is in the range of $0.75 to $3 per pound. There are few medical devices in which the raw materials cost a significant fraction of the ultimate cost to the hospital. Many of the additional charges arise from processing, insurance, and other factors involved in device production, especially packaging. Hence, one would expect that a relatively small increase in product cost would be involved even if the cost of the raw material were doubled.

An additional factor is the inherent conservatism present in the medical field. This is a well-founded conservatism and is based on the assumption that if something works reasonably well, one should not risk trying to fix it. However, when significant problems occur, the industry is certainly willing to correct those problems and introduce new materials. Hence, it needed to be shown that there was a good reason to switch over to other materials currently available. If the advantages were not great enough, even at a prototype testing stage, then no substitution occurred. Hence, a switch in materials for disposal cost reasons may actually lead to eventual improvement in actual use properties. This kind of supporting research is now needed to demonstrate the properties of these new materials in various biological settings. Finally, most of the current cost involved in producing a device is simply the up-front or production marketing cost. There is little charge involved for the disposal of a device. Hence, it is possible to transfer the cost for disposal to the user, which in many cases does not show the true cost for *not* making a substitution. What is occurring now is that the cost for the disposal is rising rapidly and the user is beginning to turn to the manufacturer to ask for assistance in reducing this disposal cost. With a clear understanding of exactly what these environmental costs are and what trade-offs are involved, a better estimate can be added to the up-front cost of the medical device, and this will also help adjust to the apparent cost advantage of PVC.

8.6 CONCLUSIONS ABOUT MEDICAL PVC AND BIOFRIENDLY MATERIALS

The substitution of PVC destined for incineration because of infectious contamination can be readily accomplished with existing resins. There may be cases in the near term where the properties of PVC need to be maintained. In these cases having PVC disposables entering the waste stream may be acceptable if these amounts are small enough so that off-site disposal is not too costly.

The search for biofriendly polymers extends well beyond finding substitutes for medical PVC to cope with infectious contamination of medical materials, the so-called regulated medical waste (RMW). Regulated medical waste represents only 20 to 25 percent of the hospital waste stream, whereas nonregulated medical waste (NRMW) represents some 75 to 80 percent. Furthermore, waste from the medical industry represents only a fraction of what is disposed of in society. In order to encourage both the development of biofriendly materials and the recycling of them, it becomes important to realize that economic value recovery is a major issue associated with achieving success. While most people agree that pollution of the environment is a real problem, few people will be willing to put forth the effort necessary to see that biofriendly materials are used unless there is value recovery associated in the recycling process.

This means that biofriendly materials should be designed with a future for them, a future that means that value recovery is possible after the item's first use. Value recovery can be divided into five sections:

1. *The biofriendly material should be recycled in its original shape.* This concept is very old and is practiced in many areas of the world, but less so here in the United States. The throwaway mentality remains dominant in the United States, and as a consequence, our waste stream is being burdened unnecessarily. Objects made of polymers should be made more durable, so that recycling in the original shape becomes a reality.

2. *The biofriendly material should be capable of reshaping.* This area offers promise for polymers since the melting point for these materials

is relatively low; thus, reshaping is energy efficient. There are opportunities to reshape polymers to new objects, but in general it means changing from a high quality object to a lower quality object. Eventually, reshaping in this manner leads to a material that must be disposed of either by biodegradation, chemical conversion, or incineration. Consequently, recycling via reshaping offers only a time delay before actual disposal becomes necessary.

3. *The biofriendly material should be capable of recycling via chemical conversion.* Again, polymers offer real opportunity here where chemistry can play an important role in the recycling aspect. Converting thermosets to thermoplastics chemically could mean that these thermosets could then be reshaped for reuse. Polymers could be designed with latent crosslinks in them such that thermosets could be converted to thermoplastics. Further, polymers could be designed with chemistry in mind which would permit their easy conversion to chemicals after the polymer's use. Thus, value could be extracted from them via the resale of chemicals made from them.

4. *The biofriendly material should be capable of recycling via biodegradation.* This approach would seem to be the most logical, yet only natural polymers really offer real promise here. Synthetic polymers are made biodegradable; however, this often means simply that their physical shape may be altered, eventually leading to powder that becomes less obtrusive visually. In fact, the material is still present in the environment; it just cannot be seen. There are only a few examples of synthetic polymers that truly biodegrade, such as polyglycolic acid and polylactic acid.

5. *The biofriendly material should be made safe for incineration.* The simplest approach to disposal of materials contaminated by infectious agents is incineration. Even when materials are not contaminated and heat value can be extracted, incineration might be the most advantageous disposal option. Of course, the potential for pollution via release of toxics into the air is quite real, and thus polymers need to be designed with better additive packages such that only water and carbon dioxide are released upon incineration. This represents a true challenge for the synthetic chemist, the material scientist, and the incinerator designer.

The final alternative to disposal of waste polymer materials is to bury them. Landfills have only a small percentage of polymers present, and if any material needs to be thrown away, plastics are among the best. They are light, and their volume is small, so they represent good candidates for landfills. However, in principal, it is best not to dispose of anything in this fashion other than that which might easily degrade once in the landfill.

Thus, the need for planning to make biofriendly materials by one of the routes mentioned above offers true opportunity for research in the future.

REFERENCES

Anonymous. 1984. The medical market. *Modern Plastics* (May): 4–5.

Anonymous. 1991. *Polymer News* 16: 25–26.

Boettner, E. A., G. L. Ball, and B. Weiss. "Combustion products from the incineration of plastics." Prepared for Solid Waste Research Laboratory, U.S. EPA National Environmental Research Center, Cincinnati, OH, 1973.

Hickling, N. 1991. *Polymer Mat. Sci. Engineer.* (ACS) 65: 285–6.

Hilado, C. J. 1990. *Flammability handbook for plastics,* 4th ed. Westport, CT: Technomic Publishing Co.

Juran, R., ed. 1990. *Modern Plastics Encyclopedia,* vol. 66.

Pasek, R. J. and D. P. Y. Chang, 1991. Potential benefits of polyvinyl chloride and polyvinylidene chloride reductions on incinerator emissions. Paper 91-33.1 presented at 84th Annual Meeting of Air and Waste Management Association, Vancouver, British Columbia, June 16–21, 1991.

Saunders, K. J. 1988. *Organic polymer chemistry,* 2nd ed. London: Chapman and Hall.

9

The Future of Medical Waste Incineration

Alex E. S. Green

9.1 STATE-OF-THE-ART ASSESSMENTS

A number of state-of-the-art reviews of medical waste management were written in 1990 and 1991 (Theodore 1990, Barton et al. 1990, Lauber and Drum 1990, Doucet 1991, Lee et al. 1991). The 1991 meeting in Vancouver of the Air and Waste Management Association (AWMA) added many others to this list, including papers that discussed general approaches (Almaula; Chang; Wollschlager and Casey; Lee and Huffman; Pasek and Chang; Schifftner; Siebert and Alston-Guiden; Rice and Hester; Jordan, Konheim and McGrane; Bulley; Green et al.). Also presented were many papers that focused on specific air pollution control devices (Hasselriis, Corbus and Kasinathan; Teller and Hsieh; Riley, Knoche and Vicinus; Maxwell; Lerner).

Comparing the number of corresponding papers in the 1990 AWMA meeting in Pittsburgh (Hasselriis; Glasser and Chang; Mineo and Rosenthal; Lauber and Drum; Green et al.), it would appear that the research and development activity in medical waste incineration is expanding rapidly. Is there a convergence of, or technical thrust emerging from, all these efforts? At this time it would appear that the conventional wisdom still focuses on the installation of air pollution control devices as the key to toxics control (see, for example, Lee et al. 1991). However, the pollution prevention or precombustion measures approach appears to be gaining credibility. As noted in Chapter 1, we may divide clean combustion technology into four general measures:

1. *Precombustion:* toxic-producing material avoidance, material selection, source separation.

170

2. *Combustion:* optimum temperature, time, turbulence, oxygen, efficient combustor arrangements and operation.
3. *Postcombustion:* heat recovery systems, scrubbers, acid gas treatment, activated carbon absorption, baghouses, electrostatic precipitators.
4. *Residue disposal:* metal extraction, solidification, roadbed use, landfill, etc.

The current trend is to add air pollution control (APC) devices to incinerators where possible or, if not possible, to use regional incinerators that have such APC devices. In contrast, in this work we have focused on precombustion and combustion measures. The practicality of such a shift is strongly dependent upon the detailed composition of the medical waste stream. Chapters 2, 4, and 5 have covered these topics. However, a very recent New York City medical waste management study provides important further details, particularly on chlorine in the medical waste stream and possible means of isolating this component.

9.2 THE NEW YORK CITY MEDICAL WASTE MANAGEMENT (NYCMWM) STUDY

Overview

Jordan, Konheim, and McGrane (1991) showed that the quantities of medical waste requiring disposal through incineration could be dramatically reduced through segregation of kitchen and food service waste, corrugated, recyclable papers, and plastic disposable medical apparatus. The operating and capital costs of this system were found to be substantially lower than other scenarios examined in the study. The segregation of plastic disposable medical apparatus and sharps with a reusable container system was found to yield the greatest cost and source reduction benefits.

Regulated Medical Waste (RMW)

The critical component of the solid waste generated by health care providers in New York City is the regulated medical waste (RMW), which has the potential for spreading infection. RMW is generally defined in the federal Medical Waste Tracking Act in New York State and City law as follows:

1. Cultures and stocks of infectious agents and associated biologicals.
2. Pathological wastes, including those from surgery and autopsy.

3. Human blood and blood products waste.
4. Used sharp implements, such as needles and syringes, and including IV bags and tubing.
5. Contaminated animal body parts.
6. Certain other surgical, autopsy, laboratory, dialysis, medical equipment, and biological wastes.

Many other items not specifically listed as RMW are usually shipped as RMW. These include, for instance, disposable medical apparatus, such as tongue depressors, gloves and gauze, which are not technically RMW, but at times may be soiled with blood or body fluids.

In the NYCMWM study representative samples were collected at 11 hospitals and sorted on-site. Product purchasing information was obtained from 38 other hospitals and correlated with this sort data to establish the composition of the waste stream for all 95 New York City hospitals. Chemical composition information was obtained from product manufacturers and from an ultimate analysis of specific materials in the New York City waste stream. The composition of the medical waste stream of acute care facilities in New York City, as established in this study, is shown in Table 9–1.

Practical Segregation Considerations

The approaches examined in the NYCMWM study for both RMW and NRMW waste streams included the substitution of durable goods for disposable items; segregation of paper, corrugated, glass, and metal cans for recycling; and the segregation of plastic disposable medical apparatus for on-site treatment and possible recycling.

Most New York City hospitals now collect patient care area waste as (1) sharps, (2) red-bag, and (3) clear-bag. Sharps and red-bag waste are both RMW. The composition of red-bag waste and clear-bag waste did not differ significantly, and items that may be defined as RMW may be found in roughly equal amounts in either waste stream. About half of RMW is "entrained" NRMW, and about half the RMW is not contained in red bags, but is "fugitive" RMW in NRMW containers.

The on-site waste sorts conducted during the study also established that almost all (94 percent) of the items defined as RMW and generated in general patient care areas are plastic. The disposable plastic items include a variety of medical apparatus, IVs, sharps, and their containers. Although sharps are collected in special puncture-proof containers, they are also found in about 10 percent of both the clear and red bags from patient care

TABLE 9–1. New York City Acute Care Facility Medical Waste Composition Total Waste Stream Components

Stream	Component	Weight % of Stream	Weight (lb/d)	Weight % of Total
RMW (as shipped)	IV bags	8.9	17,781	2.0
	Sharps	8.3	16,526	1.9
	Sharps containers	1.8	3,588	0.4
	Solution containers	8.5	16,923	1.9
	RMW plastic bags	4.4	8,776	1.0
	Animal bedding	1.2	2,502	0.3
	Apparatus	10.5	21,021	2.4
	Disposable linens	12.4	24,828	2.8
	Patient food service	10.4	20,821	2.4
	Packaging	5.4	10,859	1.2
	Paper	4.8	9,526	1.1
	Paper towels	2.9	5,751	0.7
	Other	20.6	41,310	4.7
	Total RMW	100.0	200,213	22.8
NRMW (as shipped)	Disposable linens	10.7	72,556	8.3
	Paper towels	4.2	28,548	3.3
	Office paper	5.8	39,496	4.5
	Computer paper	1.3	9,105	1.0
	NRMW plastic bags	3.8	25,993	3.0
	Batteries	0.1	381	0.0
	Animal bedding	0.3	1,793	0.2
	Patient food service	10.7	72,449	8.3
	Mixed paper	7.1	48,249	5.5
	Kitchen	15.0	101,876	11.6
	Packaging	5.2	34,967	4.0
	Corrugated	10.1	68,412	7.8
	News/Mag	6.0	40,374	4.6
	Apparatus	4.3	29,166	3.3
	Other	15.4	104,342	11.9
	Total NRMW	100.0	677,708	77.2
	Total RMW and NRMW		877,921	100.0
Summary	RMW		200,213	22.1
	NRMW		677,707	74.8
	PATHO		4,250	0.5
	RMW containers		23,691	2.6
	Summary total		905,861	100.0

Source: Waste Energy Technologies, Inc., and Konheim & Ketcham, Inc., The New York City Medical Waste Management Study, Final Report, June 1991.

areas, most often attached to IV tubing, presenting a significant risk to waste-handling workers.

The composition range of RMW from specialized areas such as laboratories and operating rooms is much broader. RMW from these areas comprises 5 to 15 percent of the RMW stream from New York City acute care facilities (ACFs). Thus, about 85 percent of the waste stream that is defined as RMW consists of plastic disposable medical apparatus and sharps from general patient care areas.

The essence of Jordan et al.'s (1991) proposal is to change medical waste management practices by adding plastic medical apparatus to the already segregated sharps stream. This would necessitate a change in the type of containers used for in-room collection. Using an enlarged sharps container would essentially eliminate the need for red bags in general patient care areas as well as laboratories and operating rooms, since virtually all waste defined as RMW from such locations would be included in this segregated stream. Combining sharps and apparatus in one container and all other waste in clear bags may contribute to more effective segregation of sharps and other RMW. Jordan et al. (1991) propose using reusable RMW containers, which would reduce the use of disposable sharps containers and red plastic bags, which make up 6.7 percent by weight of current RMW.

Nonregulated Medical Waste (NRMW)

The two major components of NRMW are kitchen and food service waste and corrugated cardboard, which are already collected as segregated streams in many hospitals. They are usually comingled at the point of disposal, the trash compactor. Recyclable paper is generated almost exclusively in administrative areas, where it makes up the major waste stream component and can be easily segregated. Glass and metal, which are generated predominantly in the kitchen and other food service areas, could be segregated and channeled into recycling markets. Corrugated boxes used in distribution, because of their size and shape, are easily segregated.

The collection and disposal of kitchen and food service waste as a separate stream was considered practical because food service waste is or can be returned to the kitchen via food service carts in most facilities, and kitchen waste is already delivered to the compactors as a separate waste stream. In summary, Jordan et al. (1991) recommend dividing NRMW into (1) kitchen and food service waste, (2) recyclable paper, (3) recyclable corrugated cardboard, and (4) recyclable glass and metals. In their view this can be accomplished with little or no disruption to current operations.

Chlorine in Medical Waste

Jordan et al. (1991) have also carried out a comprehensive analysis of the chlorine content in the medical waste from New York City's ACFs. They assume PVC to be 53.4 wt% chlorine and all IV bags and tubing to be PVC. In a survey of New York City's institutional waste stream, non-PVC plastics were found to be 0.8 percent chlorine by weight, paper (cellulosic waste) 0.27 percent, and organic waste 0.23 percent. Their results are summarized in Table 9–2. It is most interesting that IV bags and plastic gloves (in RMW and NRMW) account for about 80 percent of the 1.2 percent of chlorine in the medical waste from NYC-ACFs. The segregation of IV bags, tubing, and gloves would reduce chlorine in the remaining waste to be incinerated by 80 percent. The lime saved in the incinerators acid neutralization system (in ratio of 4 : 1 relative to chlorine per day) would result in the avoided disposal of 30,000 pounds of lime residue per day. Furthermore, since red bags for RMW collection and sharps containers are avoided, lead and cadmium emissions from pigments and plastic stabilizers will be reduced (to the extent that they are still used).

Waste and Cost Reduction

Jordan et al. (1991) note that since 1985, NYC-ACFs have experienced a 300 to 400 percent increase in the cost of their waste disposal. RMW disposal costs in 1990 was 42 cents/lb, whereas NRMW by the New York City Department of Sanitation was 2.8 cents/lb. With their proposed system for segregating of plastic medical supplies with sharps and IV bags, they could accomplish a 27 percent reduction in RMW. This, together with an aggressive recycling program for paper, corrugated cardboard, and metal and glass and an appropriate channel (composting) for food waste, would reduce the amount of hospital waste to be incinerated from 878,000 pounds to 346,000 pounds, about a 60 percent reduction.

The also consider disposal options for RMW that is presently shipped out of state. Alternatively, disposal could be made at city-financed regional facilities equipped for incineration of RMW. Finally, Jordan et al. (1991) project a capital investment need of $39,000,000 to enable NYC-ACFs to implement plastic medical in-site treatment systems. This cost would be recovered through operating savings of $21,000,000 per year. To accommodate all New York City health care providers would require a capital investment of $122,000,000 and would result in an annual savings of $53,000,000.

Application to PVC Reduction Strategy

Most of the Jordan et al. (1991) segregation strategy can be used in conjunction with the PVC substitution strategy. This change would hold open the

TABLE 9–2. Chlorine Content of New York City Acute Care Facilities Medical Waste

Component	Weight (lb/day)[a]	Plastic (lb/day)[b]	PVC (lb/day)[c]	Cellulose (lb/day)	Inorganic (lb/day)	Organic (lb/day)	Chlorine (lb/day)[d]
Segregated plastics							
IV bags	17,781	5,868	5,868	650	1,500	0	3,135
Gloves (in RMW)	3,980	3,980	1,990	0	0	0	1,078
Gloves (in NRMW)	13,476	13,476	6,738	0	0	0	3,651
Sharps	16,526	12,395	*	65	2,000	0	103
Sharps containers	3,588	3,588	0	650	350	0	31
Solution containers	16,923	11,338	0	650	150	0	93
Apparatus (in RMW)	21,021	6,247	*	1,192	3,584	0	56
Apparatus (in NRMW)	29,166	17,500	*	5,833	4,375	0	159
Subtotal	122,461			5,833	4,375		8,306
Not segregated							
Plastic bags (in RMW)	8,776	8,776	0	0	0	0	0
Plastic bags (in NRMW)	25,993	25,993	0	0	0	0	0
Other (in RMW)	37,330	21,632	*	11,980	2,066	1,652	211
Other (in NRMW)	90,866	51,216	*	30,259	5,217	4,174	505
Batteries	381	0	0	0	381	0	0
Subtotal	163,346						716

Cellulosic						
Disp. linens	97,384	29,215	0	63,300	0	405
Paper	146,750	0	0	146,750	0	396
Corrugated	68,412	0	0	68,412	0	185
Paper towels	34,299	0	0	13,720	0	37
Packaging	45,826	6,874	0	38,952	0	160
Subtotal	392,671					1,183
Organics						
Animal bedding	4,295	215	0	2,148	0	10
Kitchen/food service	195,146	29,272	31,223	48,787	1,074	535
Subtotal	199,441					545
Total	877,919					10,750
% CI Content	1.22					

[a] Includes free water.
[b] Excludes free water.
[c] Items marked (*) may contain small amounts of PVC.
[d] The chlorine content of PVC is 53.8 percent; the chlorine content of other plastics is 0.8 percent; the chlorine content of the plastic fraction of sharps containers, solution containers, and plastic bags is assumed at 0; the chlorine content of cellulosic waste is 0.27 percent; the chlorine content of inorganic waste is 0.08 percent; and the chlorine content of organic waste is 0.23 percent.

Source: Waste Energy Technologies, Inc., and Konheim & Ketcham, as cited in New York City Department of Sanitation, A Comprehensive Solid Waste Management Plan for New York City and Draft Generic Environmental Impact Statement, Appendix 4.2, March 1992.

possibility in many cases of on-site incineration of regulated medical waste, which now would consist mostly of nonchlorinated disposables. From a cost standpoint, one supplier has indicated that for rigid plastics the cost of replacing thermoform PVC sheets by PETG sheets would impose a 40 percent raw material increase, but that the disposable product increase would be only 15 to 20 percent. If flexible medical plastics can also be replaced at the same incremental cost ratio, then the saving of on-site disposal at, say, 5 cents/lb with respect to off-site disposal at 50 cents/lb should more than compensate for the slightly higher costs of the medical disposables. The prompt on-site dispositions of regulated medical waste should reduce the hazards of spreading infection by avoiding extra steps of packaging, transportation, and unloading. The fact that the institution or community generating this regulated medical waste also attends to its disposal locally should provide strong motivation to execute all procedures properly.

9.3 RETROFIT MEASURES

Many existing incinerators, when disposing of regulated medical waste having minimal chlorinated plastic, can with minor retrofitting greatly lower their emissions of various products of incomplete combustion. The retrofitting of the 1972 model Environmental Control Product Incinerator rated at 500 lb/h of No. 2 trash by the Clean Combustion Technology Laboratory provides some illustrations of the types of retrofits that can substantially improve the combustion efficiency of a relatively old incinerator at reasonable cost. While some of these modifications were carried out for the purpose of studying the cofiring of biomass with non-hazardous institutional waste they are illustrative of what might be done to an existing incinerator to extend its life and to achieve improved combustion performance.

Figure 1–3 illustrates the modifications carried out by the CCTL after restoration of the ECP. These include (1) the installation of a conveyor to load the hopper, (2) the installation of an extra blower for bottom air, (3) the installation of a secondary feeder system originally for use in a biomass cofiring program, (4) the installation of an overfire air blower originally for use in connection with the biomass feeder, (5) the installation of a reciprocating stoker, (6) the installation of various thermocouples to provide instantaneous on-line temperatures at several critical locations to a computer, (7) the installation of on-line CO, CO_2, and O_2 monitors, and (8) provisions for access ports for sampling emissions. We have also considered installing an ash pit, which would permit us to transfer the ash from a previous day's burn into an underground pit, which could be cleaned out a day later. While this ash-disposal feature would be useful for routine operation of a medical waste incinerator, it was not necessary for the CCTL experimental program. Refractory repair and refractory treatment to permit higher temperature operations was also a part of our restoration and upgrading of the ECP T500 incinerator.

The foregoing retrofits were all rather specific to the particular incinerator donated to the CCTL. As noted in Chapter 1, with relatively tight up-front management of the types of plastics in our burn bags, we achieved emission factors far lower than those achieved by typical medical waste incinerators whose emissions were measured by the California Air Resources Board. Without having undertaken specific studies of other types of 10- or 20-year-old incinerators, we cannot discuss specific retrofit measures. However, we strongly suspect that when a pollution prevention strategy is implemented, many 10-year-old incinerators can with relatively low cost achieve acceptable emission levels.

The one questionable feature of this retrofit strategy based upon the use of precombustion and combustion measures is whether the increase in particulate emissions would be acceptable. In the CCTL case the very large chimney acts in part as a settling chamber to bring us within forthcoming regulations for particulate emissions. In cases where the original installation does not have a tall stack, some hot gas particulate removal system might be needed. High temperature filter bags have been developed by the 3M Corporation for use with pulsed jet bag houses (Weber and Schelkoph 1990). If necessary in specific hospital systems, such an installation would provide the final measure needed to bring an old incinerator in good condition up to the latest particulate emission standards. The plume rise associated with hot stack gases should be a useful feature in an urban hospital (Bulley 1990), ensuring in many cases that local buildings or their air intake systems are not directly exposed to the incinerator's stack gases.

9.4 REGIONAL MEDICAL WASTE INCINERATORS

Overview

Whether to continue with the prevailing practice of on-site incineration or whether to go to a regional medical waste incinerator (ReMWI) facility is a very timely question. The California Air Resources Board has instituted dioxin regulation (see Section 1.5), which was projected to lead to the closing down of 135 of the 146 local hospital incinerators, thus almost forcing reliance on ReMWI. Lauber and Drum (1990) point out that a state-of-the-art regional biomedical waste incinerator should have negligible trace toxic emissions. A comprehensive description of the advantages of a regional medical waste incineration facility has been given by Chang (1991). We present a summary of Chang's paper.

The Regional Concept (Chang 1991)

A regional biomedical waste incineration facility accepts waste from numerous health care facilities in a geographic area. It could be centrally located near an urban area to minimize waste transportation costs or remotely located to minimize impacts on population centers and the environment. As a new facility it would naturally use state-of-the-art combustion and emissions control technology, and thus provide numerous environmental and economic benefits. First, it can replace a number of outdated incinerators at individual hospitals. A 1989 survey of 129 infectious waste incinerators at hospitals, funeral homes, laboratories, nursing homes, and veterinary offices in the state of Washington indicated that only 13 percent were installed more recently then 1986, 16 percent between 1981 and 1985, and about 32 percent during the 1970s. Forty-eight percent of these medical waste incinerators have no emission controls.

Chang argues that since many existing incinerators were installed, advances in combustion technology, air pollution control equipment, and emissions monitoring have created a new generation of incinerators that are less harmful to the environment. Furthermore, economies of scale at a larger central facility allow lower equipment and labor cost per ton of incinerator capacity. In addition, more sophisticated air pollution control equipment (e.g., dry scrubbers) that might not be cost-effective or even available in smaller sizes suitable for a single hospital incinerator can be economical for a larger facility of, say, 50 to 100 or more t/d capacity.

A typical ReMWI facility will consist of two or more combustion units to provide backup capability when one is shut down for maintenance or repairs. This also provides for more efficient operation (i.e., turndown capability) under conditions of low waste supply. The ReMWI facility will also have air pollution control equipment, a building to house this equipment and provide waste storage, and site improvements such as landscaping and roads to handle the increased traffic that it will attract.

Air Pollution Control for ReMWIs

Chang (1991) gives particular attention to the potential for installing modern air pollution control equipment in ReMWI facilities to meet current regulatory trends. There are a number of acid gas scrubbing and particulate removal technologies to choose from, each with its own technical and economic advantages, and it is likely that existing and anticipated emissions standards can be met with current state-of-the-art technology (see Chapter 6). Among the wet scrubbers described are three principal types: low energy (spray tower), medium energy (impingement type such as packed column), and high energy (venturi). Acid gas removal efficiencies of 90 percent or more can be achieved through the addition of calcium hydroxide ($Ca(OH)_2$) or caustic soda (NaOH) to the scrubber water. The cooling of the flue gases, which results from evaporation of the scrubber water, also aids in condensing and capturing heavy metals and unburned organics that were volatilized in the incinerator, as well as particulates. However to remove fine particulates, fan horsepowers can be as high as four times that of fabric filter systems. Furthermore, wet scrubbers are inherently subject to corrosion and erosion problems, require disposal of scrubber waste liquids, and can produce a visible steam plume.

Chang also describes dry scrubbers that use dry powdered lime instead of liquids to absorb acid gases. He notes, however, that while dry scrubbers are currently economical for use in large municipal waste incineration systems, their chemical costs and space requirements may not be advantageous for medical waste facilities. On the other hand, wet/dry scrubbers, sometimes called *spray dryers,* in which a lime slurry is substituted for dry lime might provide a good compromise, and the particulates and unreacted lime would be collected in a baghouse downstream of the scrubber.

Traffic, Building, Screening, and Landscaping

Infectious and pathological wastes when sent to ReMWI are typically packed in plastic-lined corrugated cardboard cartons and shipped via tractor-trailer rigs. General hospital waste is usually shipped in compactor containers or *rolloff* boxes on special chassis. Since the vehicles arriving at the facility will be heavy trucks typically found on highways or in urban settings, access to highways and major arterial roads is important. Surface streets must be adequately designed for the weight and easy maneuvering of tractor-trailers. From a public relations standpoint, it is best to avoid truck routes that pass in front of schools, parks, and playgrounds. On-site roads must be adequate to queue all incoming vehicles to avoid blocking neighborhood streets as they wait to unload (Chang 1991).

Waste-unloading activities at a ReMWI facility must take place within a totally enclosed building for aesthetic reasons, to minimize the entry of birds and insects, and to contain any noise, dust, and odors. In milder climates, the incinerators, boilers, and air pollution control equipment may all be located outdoors, saving on building construction. In northerly climates, this equipment should be located indoors, especially when acid gas scrubbing systems involve lime handling, which is sensitive to humidity and freezing conditions (Chang 1991).

Some states and local health agencies limit the time that biomedical waste can be stored at ambient temperatures. Thus, it is prudent to provide refrigerated storage capability for the infectious portion of the biomedical waste stream, or to have emergency disposal agreements with commercial biomedical waste facilities or hospitals with in-house incinerators or large capacity steam sterilizers (autoclaves) in the event of major equipment outages.

Because of the public's general resistance to all solid waste facilities, landscaping to screen or hide the facility will be necessary. Effective landscaping can include berms as well as trees and shrubs that provide year-round screening.

Sponsorship of ReMWI Facilities

The sponsors who are developing or using ReMWI facilities include (1) hospital associations, consisting of a regional group of hospitals and medical facilities who have banded together to address problems common to health care facilities, and (2) waste management firms, consisting of businesses formed to transport and incinerate infectious and general waste from hospitals, medical and dental clinics, research laboratories, funeral homes, and veterinary clinics whose individual waste streams may be too small to be handled independently.

Chang also notes that steam sterilization of biomedical wastes in large autoclaves or retorts is a well-established technology permitted under current regulations. Most medical facilities treat at least some of their infectious wastes via on-site steam sterilization. Regional stream sterilization facilities (alternatives to ReMWIs) in California and other parts of the country are generally owned and operated by the private sector.

In summary, Chang concludes that (1) a regional medical waste incineration facility can replace a number of smaller, outdated hospital incinerators, (2) by using state-of-the-art combustion and emissions control equipment, ReMWI facilities have the potential to achieve a new improvement in regional air quality, and (3) hospital associations and commercial waste management firms are viable sponsors for regional incineration facilities.

9.5 THE FUTURE MEDICAL WASTE INCINERATOR

Section 9.3 described some simple retrofit measures to improve combustion efficiency, which in conjunction with a toxic minimization program can lower the emission levels of many existing incinerators to within acceptable levels. Section 9.4 described the ReMWI approach, which does not assume precombustion measures but instead relies primarily on good combustion and modern postcombustion systems. In this section, we address the types of combustion systems now or soon to be available for on-site or regional applications.

An incinerator wholly designed and manufactured in New Zealand was described in Chapter 3 and illustrated in Figure 3–2. This is essentially a three-stage unit, which achieves a residency time of 1.5 seconds and an afterburner temperature of more than 1,050°C (1,925°F). From its low dioxin and furan emissions (see Figures 3–3 through 3–6 and 3–8 through 3–12) this is clearly a state-of-the-art incinerator.

The low chlorinated organic emissions of the Clean Combustion Technology incinerator, which is a retrofitted 1972 Environmental Control Products T500 incinerator (see Figure 1–3), suggests that it might now qualify as a state-of-the-art incinerator. The major mechanical upgrades included extra bottom air and overfire air blowers, the stoker, and biomass feeder. As described in Section 1.4 a data acquisition system has been installed that provides O_2, CO_2, and CO readings and records on a PC various control settings every 11 seconds. An effort is under way (Wagner 1992) to use the on-line information to assess the state of combustion and to activate feedback controls to minimize products of incomplete combustion (PICs).

Among current commercial units, a pulsed hearth system clearly stands out as a state-of-the-art incinerator (Basic 1990). This incinerator has a U-shaped hearth suspended by cables, which is pulsed intermittently by air bags to agitate the waste to improve combustion. To maximize turbulence, combustion air is provided by thermal exciters in the second and third stages in the form of a series of jets. A fourth stage combines recirculated gases from stage 3 to control exhaust temperatures to avoid slagging.

Rotary kilns, which have been used extensively in hazardous waste incineration, also can be used for disposing of medical waste. The agitation provided by the tumbling motion leads to high combustion efficiency. An equivalent agitation of the waste can be achieved with a water-cooled auger made from high temperature alloy. Feeding systems are available that effectively avoid the temperature surges characteristic of batch-fed incinerators.

In some standard commercial systems, transfer rams are used with stepped hearths to agitate the waste in the primary chamber. Properly supplied with air, these can lead to high primary chamber combustion efficiencies. Many other feeding systems–primary chamber arrangements have been developed and are provided by incinerator manufacturers to achieve good primary stage combustion efficiency. A recent survey of commercial units has been given by Barton et al. (1990). Fluidized bed technology, originally developed for coal burning to minimize NO_x and SO_2 emissions, is also being considered for medical waste incineration.

The relatively recent tendency to install air pollution control devices has added a new dimension, which is strongly influencing the design of the future medical waste incinerator. One suspects that the use of starved air, which was primarily used to minimize particle agitation, will be phased out since some form of postcombustion control will mitigate the particle problem. Since starved-air operation generates many products of incomplete combustion, this mode is undesirable from the toxic minimization standpoint.

The impact of front-end avoidance of chlorinated organic inputs, toxic metals, and other toxic-producing materials by rigorous purchasing specification and segregation procedures has scarcely been examined. Thus we are not yet ready for a shootout between various types or arrangements of medical combustion systems. Without doubt the envelope of acceptable operational performance will be greatly broadened by the pollution prevention approach. Which of the competing technological arrangements will simultaneously minimize costs and environmental emissions remains for future competitions to decide.

The current tendency is to incorporate in new incinerator designs the regulatory rule of exposing primary combustion chamber (PCC) exhaust gases to 1,800°F for 1 second in the secondary combustion chamber (SCC). From the physical or chemical viewpoint of good toxic destruction, this rule is rather arbitrary. Cassito (1985) has provided temperatures to achieve 99.99 percent (4-nines) destruction of the very refractory chlorinated organics (2,3,7,8-TCDD) at temperatures from 1,340 to 2,240°F. These data can be fit by the rule (Wagner (1992):

$$t = 1.41 \times 10^{10} \exp - T/77$$

where t is time in seconds and T is temperature in °F. This equation can be solved for the temperature needed for a specific residence time:

$$T = 1800 - 77 \ln t$$

At 1,800°F, 1.0 s is required for 4-nines destruction. At 1,900°F, only 0.27 s is needed. For a half-second residence time, the temperature need only be raised to 1,853°F. A time-versus-temperature regulatory rule would allow older hospital incinerators with small afterburners to continue to operate, provided the last combustion chamber's minimum temperature can be increased to meet this rule.

Hasselriis (1992) addressed the principles of combustion control and shows how automatic controls, with the understanding and assistance of the operator, can operate a starved-air system at its optimum charging rate without exceeding its combustion capacity. He presents data from operation of medical waste and municipal waste incinerators, which indicate that CO levels less than 10 ppm can be maintained within a wide range of temperatures provided that sufficient oxygen is ensured.

Kuntz and Gitman (1990) of American Combustion, Inc., have reported the use of oxygen-enhanced combustion technologies in hazardous waste incineration. The possibility of using oxygen enhancement to meet special conditions of medical waste would appear to warrant investigation.

The technology of high destruction efficiency combustion is represented by a large literature on toxic and hazardous waste incineration (Oppelt 1987) growing out of our multibillion dollar Resource Conservation and Recovery Act (RCRA) or superfund programs. This literature, of course, can be applied to medical waste incineration to the extent that any residual toxic-producing materials are represented in the waste stream after a pollution-prevention approach is adopted.

9.6 BROADER EMISSION ISSUES

The Continued Role of Fire

Fire has been a civilizing force for humankind since the dawn of civilization, sometimes compared with the discovery of speech, writing, and agriculture. However, carbon dioxide, pollutants, and toxic emissions from fire are now viewed by some as a threat to civilization. Figure 1–2 provides an overview of local, regional, and global anthropogenic emissions to the atmosphere largely associated with the use of chemical fire for energy purposes. Which of these problems is the most urgent at this time is a matter of debate. This work is primarily concerned with the issue of toxics from medical waste incinerators. However, as we shall show, there are close relationships between this problem and a number of broader anthropogenic emission issues.

Municipal Waste Incinerators and Their Emissions

Municipal waste is now regarded as a significant source of fuel for electricity generation. Figure 9–1 gives the breakdown by weight of the 1988

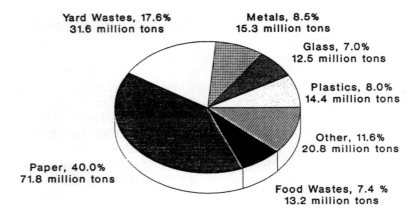

TOTAL WEIGHT = 179.6 million tons

FIGURE 9–1. Materials generated in MSW by weight, 1988. (From U.S. EPA 1990.)

municipal waste output in the United States (U.S. EPA 1990). Of the approximately 180 million tons generated, one might conservatively project the use of 100 million tons for waste to energy use. Granting an average of 5,000 Btu/lb, one readily calculates the energy potential

$$10^8 \text{ t} \times 2.10^3 \frac{\text{lb}}{\text{t}} \times 5.10^3 \frac{\text{Btu}}{\text{lb}} = 10^{15} \text{ Btu} = 1 \text{ quad}$$

If one adds waste biomass from forestry operations or from agriculture, this source of energy might readily reach about 3 quads. Cultivated biomass energy crops could provide some 8 extra quads according to the pessimists or as much as 12 quads according to the optimists. In either event, the sum would be a significant part of our national energy use, which currently totals about 80 quads.

The combustion of municipal waste and biomass, of course, will contribute CO_2 to the atmosphere and to this extent would exacerbate the greenhouse problem. However, for disposables such as newspaper made from biomass, the uptake of CO_2 in growing the pulp would largely balance the emissions of CO_2 during combustion. Accordingly, from the CO_2 standpoint the use of energy from the combustion of biomass and much of municipal waste is a natural component of a solution to the greenhouse problem.

Some of the other gaseous emissions of municipal waste facilities also absorb in the long wave earth radiation region, which controls the global atmospheric temperature. The 800 to 1,200 cm^{-1} region or 12.5 to

8.3 μm region is of particular importance to the anthropogenic greenhouse problem, since this region is near the peak of the earth-atmosphere long wave radiation that is not blocked by H_2O and CO_2 bands. Many PICs absorb in this region, so the eventual use of municipal waste, biomass, and fossil fuels must be considered carefully in this context. Figure 9–2 shows the black body spectrum for 255 K, which calculates to be the average global temperature of the earth-atmospheric system in the absence of greenhouse gases. Also shown are the active absorption bands of H_2O, CO_2, O_3, NH_3, CO, CH_4, NO_2, NO, SO_2, and liquid water and the positions and approximate band strengths of important chlorofluorocarbons (CFCs) in the long wavelength region (Green 1989b).

Table 9–3 is a list of the spectral features of C_1 to C_3 products of combustion including chlorinated PICs, which have emissions or absorption bands in this range (Hall, Lucas, and Koshland 1991). Thus, PICs might play a significant role in the anthropogenic greenhouse problem. Table 9–4 presents some data on long-lived industrial solvents that are also PICs. Most of these lifetimes are determined by reactions with OH. Emissions of CO, a major product of incomplete combustion, can lower the natural OH concentrations, which would lengthen the lifetimes and hence increase the equilibrium concentrations of the greenhouse gases in Table 9–4. Not enough attention has been given to the relationship between PICs, particularly chlorinated PICs, and the greenhouse problem.

Long-lived chlorinated PICs should diffuse into the stratosphere, where they would be photodissociated by the harder ultraviolet radiation of the sun. In these cases the chlorine atoms created can catalytically destroy ozone molecules. Thus, chlorinated PICs such as are emitted from medical waste are greenhouse and ozone depletion threats.

The larger municipal incinerators when properly operated expose chlorinated PICs to long times at high temperatures, which helps break down these compounds. However, institutional combustion facilities, particularly medical waste facilities, are much smaller and when operated in the conventional starved-air mode can be a large source of toxic PICs as well as greenhouse and stratospheric ozone-depleting gases. Burning disposable chlorinated plastics poses a triple threat to society: toxics, ozone depletion, and greenhouse enhancement.

9.7 DIVERGENT VIEWS ON WASTE INCINERATION

The MIT-CTPID Study

Ehrenfeld et al. (1988) of the Massachusetts Institute of Technology (MIT) Center for Technology Policy and Industrial Development (CTPID) have

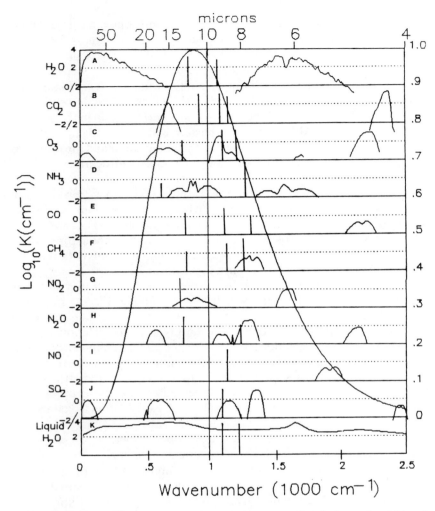

FIGURE 9–2. Greenhouse spectral region. The large curve in relation to the right scale represents the earth's relative long wave spectral radiance versus wave number (lower scale) and wavelength (upper scale). The smaller curves for the trace gases in the 10 strips give $\log_{10}k$, where k is a moderate resolution molecular band absorption coefficient. The lowest strip gives $\log_{10}k$, where k is the liquid water Lambert-Beer absorption coefficient. Both H_2O strips have vertical ranges of 0 to 4, whereas all others range from -2 to $+2$. The straight vertical lines in the strips labeled A, B, C, . . . give the central locations of CFC's bands with heights represented by $\log_{10}I$, where I are the integrated band intensities all on vertical scales from 0 to 4. A = $CFCl_3$ (F11); B = CF_2Cl_2 (F12); C = CF_3Cl (C13); D = CF_4 (F14); E = $CHClF_2$ (F22); F = C_2F_6 (F116); G = CCl_4; H = $CHCl_3$; I = CHF_3; J = CH_2F_2; K = $CBrF_3$. (Adapted from Green 1989b.)

TABLE 9–3. Spectral Features of Products of Incomplete Combustion in Earth's Long Wave Radiation Region

Wavelength (cm^{-1})	Species
793.9	Carbon tetrachloride
795.8	1,1-Dichloroethylene
801.4	Tetrachloroethylene
824.6	Ethane
826	*trans*-1,2-Dichloroethylene
850	Trichloroethylene
857	*cis*-1,2-Dichloroethylene
858.5	*trans*-1,2-Dichloroethylene
869	1,1-Dichloroethylene
896.6	Vinyl chloride
898.6	*trans*-1,2-Dichloroethylene
917	Tetrachloroethylene
937.3	1,1,2-Trichloroethane
940	Trichloroethylene
942.3	Vinyl chloride
949.6	Ethylene
973.2	Ethyl chloride
1,020.2	Methyl chloride
1,033.5	Formaldehyde
1,086.2	1,1-Dichloroethylene
1,087.4	1,1,1-Trichloroethane
1,095.2	1,1-Dichloroethylene
1,097.4	1,1-Dichloroethylene
1,200.6	*trans*-1,2-Dichloroethylene
1,209.3	1,1,2-Trichloroethane
1,220.7	Chloroform
1,232.2	Ethylene dichloride

Source: Adapted from Hall, Lucas, and Koshland 1991.

analyzed advocacy positions relative to hazardous waste incineration and municipal solid waste incineration. While there are significant differences between these two issues and both differ with respect to medical waste incineration, many of the CTPID considerations are, from this writer's observations, also reflective of the divergent views on medical waste management. The CTPID have thoroughly confronted the sources of disagreement in waste incineration disputes and note that despite the efforts of many groups, there has been little movement towards the solution of waste management problems. Since inaction in waste management poses many social problems, the MIT-CTPID undertook a study, the goal of which was to define the critical issues in disputes, identify the major players, and assemble representative positions in a format that would clarify the positions for all parties. The CTPIC began discussions at a

TABLE 9–4. Products of Incomplete Combustion That Are Long-Lived in the Atmosphere

Compound	b.p. (°C)	Half-life	Decomposition
Benzene	118.5	26.7 days	OH reaction
Carbon tetrachloride	76.5	30–50 years	OH, photolysis
Chloroform	61.7	80 days	OH reaction
1,1-Dichloroethane	57.3	62 days	OH reaction
Pentachloroethane	161	1.8 years	OH, photolysis
Tetrachloroethylene	121	2 months	OH reaction
Toluene	110.6	1 day	OH reaction
1,1,1-Trichloroethane	74.1	0.5–25 years	OH, photolysis
Trichlorofluoromethane	23.7	52–207 years	Photolysis

Source: From Green et al. 1990a.

conference that allowed the participants to converse and share each other's concerns.

Methods of CTPID Study

Characterization of the Current Dialog
A representative sample of the recent scientific literature, interest group position papers, testimony, and articles, as well as regulatory reports connected with hazardous or municipal waste combustion was assembled.

Interviews
Parties associated with a proposal to build a hazardous waste incinerator in Braintree, Massachusetts, were interviewed, and 10 in-depth interviews with project proponents and opponents, as well as local government officials, were conducted.

Data Analysis
Data analysis comprised three steps: (1) defining the range of issues in dispute, (2) identifying the positions taken by the interest groups within each issue, and (3) designing a scheme to classify the positions.

Preconference Surveys
Conference participants were given the CTPID initial position paper 1 month in advance.

Conference Discussion Session
The first substantive discussion session at the conference was devoted to this paper.

Postconference Revisions
After the conference, the CTPID revised the paper to reflect the participants' suggestions.

Factors That Sustain the Conflict

The CTPID study found that 5 factors sustained the incineration conflict:

1. The following three factors influence nearly all the separate issues and positions:

   a. *Need:* This is at the heart of many incineration debates. Opponents contend that incineration will create disincentives for waste reduction and recycling. Proponents argue that existing facilities are not adequate and incineration is urgently needed.

   b. *Safety and reliability:* In the absence of definitive scientific data, parties argue over the risk of chronic disease from stack emissions and ash disposal. How does the risk compare to other risks we all encounter every day? Will the facilities, once in place, operate as intended? Will they always meet all the safety standards? How can government and the public keep the operators honest and on their toes?

   c. *Fairness:* Is it fair to ask one community to host an incinerator that treats wastes from elsewhere? Are siting laws and the consequent procedures fair to all parties—neighbors, developers, workers, others?

   To be acceptable to all concerned, decisions about incineration must address each of the three above-mentioned concerns. Agreements must carefully weave together considerations in all of the three areas, recognizing some overlap and involvement of economic issues in each.
2. In public debates, parties often tend to express their positions in absolute terms. Such absolute conflicting views of the world stand in the way of reaching mutually acceptable solutions. Parties draw their own conclusions from scientific evidence.
3. Parties perceive the dispute as a battle in which a winner and a loser will be designated.
4. People want certainty from science, but science cannot provide certainty about many of the key questions within the incineration dialog. Scientific findings and social declarations of values often are confused.
5. Once parties formulate their positions, their willingness to assimilate and consider contradicting information diminishes. The learning process slows down.

Parties Involved

The major parties involved in incineration disputes include the following:

- *Citizen groups:* primarily local groups formed to oppose particular incineration projects
- *Environmental groups:* nationwide, regional, local groups and private organizations generally involved with broader environmental issues
- *Industry:* hazardous waste generators, waste transporters, waste services contractors
- *Government:* federal, state, and local agencies and organizations of government officials
- *Professional societies:* national organizations of professionals, e.g., ASME, AWMA, and so forth

Views of Various Groups

The views of the five above-mentioned groups on the three issues of need, safety and reliability, and fairness are summarized in Tables 9–5, 9–6, and 9–7. These tables may be taken as the distillate of the MIT-CTPID study of advocacy positions on hazardous waste incinerators and municipal waste incinerators. As should be obvious, there are almost irreconcilable differences on these issues among the five groups. Most of these same views in all likelihood prevail on issues involved in siting of a medical waste incinerator. However, here one must identify at least one additional group, whose views strongly influence the decision-making process, namely, hospital administrators.

Views of Hospital Administrators

Recycling or composting is not feasible for infectious components of the medical waste stream, i.e., regulated medical waste (RMW). Thus there clearly is a need for medical waste incineration. Bulley (see Section 3.9) has discussed alternatives to incineration: (1) landfilling, (2) autoclaving and then landfilling, (3) microwaving and then landfilling, and (4) hammermilling and chlorinating with the liquid effluents sent to the sewer and solids to the landfill. None of these alternatives destroys toxics. Furthermore, the intermedia transfer of toxics or infectious material by these alternatives can create greater problems than they solve. Thus the need and the safety and reliability issues raise the following questions:

1. *On-site incineration:* Retrofit existing incinerators (see Section 9.3) or purchase a new one (see Section 9.5)?

TABLE 9-5. Need

Citizen
a. Safe, workable alternatives to incineration exist, but government and industry are not interested.
b. Solid waste incineration rules out comprehensive waste reduction, recycling, and composting.

Environmental
a. The nation is facing a solid waste generation and disposal crisis.
b. Incineration cannot be considered apart from a comprehensive waste plan, which includes waste reduction and recycling.
c. The need for incineration will be reduced or eventually eliminated by sensible economic and health-related measures.

Industry
a. Additional treatment and disposal capacity is needed now.
b. Landfilling is becoming increasingly unattractive and unavailable. Federal regulations prohibit landfilling of many wastes.
c. Waste minimization is not enough; additional treatment and disposal facilities are necessary.
d. Incineration is an appropriate technology for waste treatment.

Government
a. Improper waste treatment and disposal endangers the public's health and safety. Alternatives are needed.
b. Demand for incineration will increase as landfills reach capacity and federal restrictions on land disposal are implemented.
c. Managing wastes requires a comprehensive approach. Waste reduction and recycling are essential but not sufficient.

Professional Societies
a. Incineration is a necessary technology to address the nation's hazardous and municipal waste disposal problems.

Source: Adapted from Ehrenfeld et al. 1988.

2. *Regional incineration:* Organize by hospital consortium or commercial service?

The pros of regional incineration were discussed in Section 9.4. Some of the cons include the following:

1. Is the extra cost of regional incineration at, say, 50 cents/lb justified with respect to on-site incineration at, say, 5 cents/lb for an amortized incinerator or 10 cents/lb for a new on-site incinerator?
2. What extra packaging costs (labor and materials) are needed for off-site shipment of RMW?
3. What are the potential liabilities to the hospital associated with improper disposal by others in off-site incinerators?

TABLE 9–6. Safety and Reliability

Citizen
a. People have despoiled the earth with toxic chemicals.
b. Incineration is an unsafe technology.
c. Incineration is not a disposal technology. Incineration reduces the volume of solid waste, but enhances its toxicity.
d. Citizens are skeptical of assurances from government and industry.
e. The proposed site is unsuitable.

Environmental
a. No incinerators should be built for at least the next 5 years.
b. If uncontrolled, incinerator air emissions are hazardous to human health.
c. Stringent emission controls are necessary.
d. Incineration should not be used for wastes that create hazardous air emissions or ash.
e. Municipal solid waste incinerator ash can qualify as a hazardous waste; it should be tested and disposed of accordingly.
f. Operator training/certification programs are necessary to ensure proper facility operation.

Industry
a. Federal and state governments impose unnecessarily strict standards on hazardous waste incinerator operations.
b. Hazardous waste incinerator emissions do not pose a threat to human health or the environment.
c. The public's safety concerns are based on emotions, not facts.
d. Municipal waste incinerator ash can be disposed of safely without treating it as a hazardous waste.

Government
a. Our knowledge regarding incineration of waste is incomplete.
b. Uncertainty must not stand in the way of action to address the management problem.
c. Hazardous waste incineration poses lower risks to health than alternatives.
d. Atmospheric emissions from properly designed and operated incinerators pose acceptable health risks.

Professional Societies
a. Incinerator emissions are not a threat to health or the environment.
b. Operator training and facility maintenance requirements are best addressed by industry or professional organizations.

Source: Adapted from Ehrenfeld et al. 1988.

On the fairness issue, it would seem that exposing a community in the neighborhood of a regional incinerator to the emissions and associated risk from the burning of medical waste from many communities is intrinsically unfair. Since a community benefits from the health service provided by local hospitals and health care facilities, it should accept the risks associated with these services. If it does so intelligently, it will insist that disposal be done carefully to minimize risk. One might debate this conclusion if the regional incinerator achieves, say, 6-nines destruction and the local

TABLE 9-7. Fairness

Citizen
a. Government and industry are dictating waste policy.
b. Siting decisions are based on politics, not science.
c. It's unfair to ask one community to bear the risk of a societal problem.

Environmental
a. Host communities should be compensated for bearing the risk of an incinerator.
b. Citizens must be assisted and encouraged to participate as equals early in the decision-making process.

Industry
a. The public can benefit from incinerator operations.
b. Governmental regulations can be overly restrictive.
c. The public must share the responsibility for waste disposal.

Government
a. Programs for public participation are in place; the current system ensures democratic, fair, decision making.
b. Managing solid and hazardous wastes is everyone's responsibility. (When government acts in accordance with this responsibility, citizens may perceive its actions as unfair.)
c. The federal government has a responsibility to ensure the consistency, and thus the fairness, of disposal practices nationwide.

Source: Adapted from Ehrenfeld et al. 1988.

incinerator achieves only 4-nines destruction. At this time such debates would involve great uncertainties and would get into technical issues upon which professionals may differ and that are certainly beyond the scope of the lay public.

In actuality, decisions by hospital administrators on how to meet soon-to-be-implemented regulations will be strongly influenced by local questions, such as:

1. Is the incinerator in a residential neighborhood, industrial park, open area, low population density suburb, or outskirts?
2. Is there enough space for retrofitting or a new incinerator?
3. If we install a new system at a nonconspicuous local site, could we make money by providing service to other community health providers who are now subscribing to a commercial service?
4. Would a small hospital subscribing to a larger hospital's disposal service potentially lose a local competitive edge?
5. Is there a negative public perception of the hospital that conducts on-site incineration?
6. Is it prudent to be a leader in adopting a pollution prevention approach, or is it safer to wait until the dust settles?

Views of Professionals

Aronson and Hasselriis (1991) have identified many current public misperceptions. One example is the notion that heavy metals are a problem when waste is incinerated but not when it is composted. Actually the leachable metals are similar in both cases and may or may not be a problem. Reconciling such public perceptions with reality will require a suspension of the usual confrontational approach to such issues, a substantial research and development effort to resolve technical uncertainties, and a major public education effort.

9.8 RECONCILIATION BY POLLUTION PREVENTION

The divergent views on municipal waste incineration and hazardous waste incineration categorized in Section 9.7 appear almost irreconcilable. Yet in the case of medical waste incineration or, more generally, institutional waste incineration, there is hope. The possibility of strong control of the input waste stream is the major feature that differentiates institutional waste management from municipal waste management and hazardous waste management, which have had almost all of the attention of federal and state programs (Green et al. 1990). With the guidance of the chief executive officer (CEO), an institution's purchasing agent can restrict the input into the institution's waste stream so as to minimize toxic products of waste combustion. An institution can also develop source separation methods and motivational techniques to limit the toxic material or toxic-producing material entering its burn bags. In particular, restricting halogenated organic compounds and toxic metals in the input to the institution and in the burn bags could substantially lower the production of corrosive and toxic products. A regional medical waste incinerator (MWI) or municipal solid waste incinerator (MSW) involved with much more diverse sources of waste, would have more difficulty developing preventive and motivational methods to lower the toxics production although there are possibilities (Green et al. 1992). The option of controlling the toxic input is not available to hazardous waste combustion facilities, whose very purpose is to dispose of toxic wastes.

Most assessments of institutional, particularly hospital, waste combustion draw primarily upon experience with hazardous waste combustion and only consider combustion and postcombustion measures. They neglect pollution prevention or precombustion measures, the first law of modern fuels and combustion technology (FACT). This lack of consideration of restricting or cleaning the input to the institutional waste stream

TABLE 9–8. Clean Combustion Technology for Medical Waste

Precombustion

1. Send cytotoxic, hazardous, and radioactive chemicals to toxic depository.
2. Replace disposable toxic material (TM) and toxic producing material (TPM), by non-TM and non-TPM.
3. Recycle reusables after sterilization.
4. Use motivational systems to ensure sanctity of burn bags.

Combustion

1. Coburn with natural gas as needed.
2. Use stoker, moving grates, auger, pulsed hearth, or rotary kiln.
3. Afterburn (reburn) with natural gas.
4. Optimize temperatures and CO levels.

Post Combustion (as needed)

1. Hot gas cleanup.
2. Spray drier and fabric filter.
3. Venturi scrubber and packed tower.

Residue Disposal

1. Extract metals from ash.
2. Immobilize.
3. Landfill (monofill if necessary).

imposes great limitations on possible solutions to the problems of corrosive or toxic products from institutional combustors. In contrast, the multibillion dollar Clean Coal Technology Program has given considerable attention to the development of coal cleaning technologies. Using low sulfur coal by selective purchasing is also a common practice in coal-burning technology to avoid combustor degradation and costly postcombustion measures or to lower the cost of such measures. In municipal waste systems the refuse derived fuel (RDF) approach (Saltiel 1991) involves precombustion processing intended in large part to clean up the input waste stream. Recycling paper, plastics, glass, and metals in conjunction with municipal waste disposal is an alternative precombustion measure that can minimize combustion and postcombustion problems.

In the case of institutional incinerators, this work indicates that restricting the use of PVC and toxic metal pigments in the disposal waste stream would greatly reduce corrosive and toxic products of combustion. In this case combustion can continue to serve its principal roles of destroying infectious agents, reducing the volume and weight of the waste, and replacing fossil fuel as a source of heat or other forms of energy. Table 9–8 summarizes what might be called the three laws (or measures) of clean combustion technology as applied to medical waste incineration. A regulatory approach, in which emission limits are specified but the institution has its choice of measures to achieve these limits, would undoubtedly stimulate cost-effective approaches using combinations of precombustion, combustion, and postcombustion measures.

Prevention vs. Cure

As noted in Section 9.1, most recent state-of-the-art assessments of medical waste incineration in America focus on the use of air pollution control (APC) as the key to advances of medical waste incineration (see, for example, Lee, Huffman, and Nalesnik 1991; Barton et al. 1990; and many others). On the other hand, the thrust of this work is elegantly summarized in the old adage "An ounce of prevention is worth a pound of cure." The specific essence of the pollution prevention (PP) approach to medical waste incineration is compactly stated by Bulley (1991b):

> Summary: The replacement or substitution of halogenated plastics and the removal of toxic heavy metals which are used as fillers, stabilisers, and pigments is feasible. This change can have a significant effect on the environmental impact that a medical waste incinerator has on its environment.
> Conclusions: A policy of giving purchasing preference wherever possible to nonhalogenated plastics and plastics which do not contain toxic heavy metals is not only feasible but will also result in: a marked reduction in emissions of hydrogen chloride, a consequential reduction in the amount of toxic dioxins and furans generated as products of incomplete combustion, and a marked reduction of heavy metals in both the stack off-gases and in the residual ash.

Figure 9–3 provides a flow diagram for waste at a large hospital based upon Clean Combustion Technology Laboratory experiences described in Chapter 1 and the analysis of Jordan et al. (1991) as modified to incorporate the substitution of PVC by nonchlorinated plastics. An inexpensive fly ash collection system such as the core separator–cyclone device (Smolensky et al. 1991) could cap off the system.

Reconciliation of Divergent Views by
Pollution Prevention

The wide divergence of views of government, industry, professional societies, environmental groups, and citizens groups described in Section 9.7 might conceivably be reconciled by the general worldwide adoption of a broadly formulated pollution prevention ethic. As noted in the preface of this book and in Section 1.1, pollution prevention programs have already

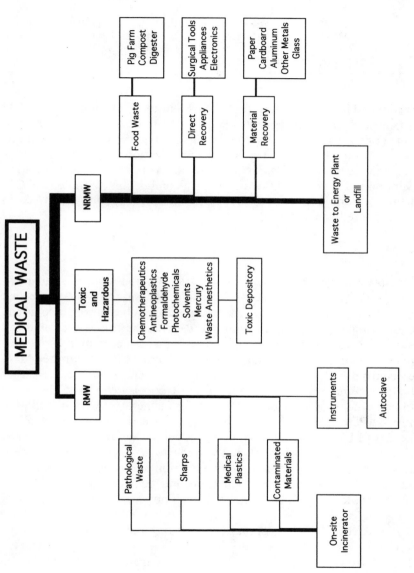

FIGURE 9-3. Flow diagram for medical waste.

TABLE 9–9. Approximate Emission Factors for Criteria Pollutants (ppm)

	HCl	PM	CO	NO_x	SO_2
U	1[4]	1[3]	1[2]	2[3]	1[3]
C	5[2]	1[1]	1[2]	1[3]	1[2]
PU	~1[3]	~5[2]	~1[1]	~1[3]	~5[2]
PC	~1[2]	~1[1]	~1[1]	~5[2]	~1[2]

been initiated in the United States by the Pollution Prevention Act of 1990 and as described in writings by Deland of the Council of Environmental Quality (1991). The focus, however, has mainly been on recycling, reduced packaging, and other methods of waste minimization in specific industries (Licis et al. 1991). A major aim of this work has been to extend pollution prevention to include the first law of clean combustion technology (CCT) and to apply the combined thrust of pollution prevention and CCT to medical waste incineration.

Recent Emission Factors

The simplest scheme for estimating ballpark toxic emissions from an incinerator is based upon the use of measured emission factors. Here one reports the output weight per hour of a specific pollutant or toxic as the fraction of the total weight per hour of the input medical waste. Table 9–9 lists emission factors for criteria pollutants in parts per million by weight. The second row gives approximate emission factors for uncontrolled MWI selected from an assembly of cases that gave low emission factors (Lerner 1991; Teller and Hsieh 1991; Walker and Cooper 1991). The third row gives corresponding low emission factors for incinerators with air pollution control (APC) devices. The fourth row gives estimates of what might be achieved by reasonable pollution prevention measures following the practices illustrated in Figure 9–3. The fifth row gives estimates as to what might be achieved by combining good pollution prevention measures with good, but not overly costly, air pollution control measures. To emphasize the fact that the science of emission factors is not even good to one decimal place, we give all numbers as 1, 2, or 5 with the power of ten in square parenthesis. Thus 5[2] stands for 500 ppm. To emphasize the speculative nature of rows 4 and 5, we insert ~ marks.

Table 9–10 gives emission factors for heavy metals in parts per billion (ppb) by weight. Here the estimates have even greater uncertainties. Despite these caveats, we believe Tables 9–9 and 9–10 provide a reasonable summary of the current state of the field and what might be achieved in the near future by pollution prevention.

TABLE 9–10. Estimated Emission Factors for Heavy Metals (ppb)

	As	Cd	Cr(T)	Cr(6)	Hg	Mn	Ni	Pb
U	5	5[2]	2[2]	2[1]	1[2]	1[2]	1[2]	1[3]
C	2	5	2[1]	1[1]	2[2]	2	2	2[1]
PU	~1	~2[1]	~2[1]	~5	~1[1]	~1[1]	~1[1]	~2[1]
PC	~2[−1]	~5	~5	~1	~2	~2	~2	~5

As an illustration of the potential application of Tables 9–9 and 9–10, we might consider a community that is faced with a decision as to whether to permit a regional MWI within its boundaries. Let us say the community hospital's generation of regulated medical waste (RMW) is 200 lb/h. To estimate the community's own fair allocation of the various emissions, we would multiply its generation rate by the U or C or PU or PC emission factors. Now let us suppose a regional medical waste incinerator is proposed, which intends to burn 2,000 lb/h of red-bag waste. To ensure that the local community is not exposed to any more toxics than it would generate itself, the technology of the regional incinerator should have emission factors $1/10$ of those deemed reasonably attainable by a community MWI. These simple calculations do not take into account dispersion aspects that are dependent upon local wind velocities, directions, and stability conditions. In this latter connection it might be noted that if fog conditions occur, most pollutant dispersion models break down and ambient concentrations in the incinerator's neighborhood could build up to a fraction like 0.1 or 0.01 of the concentrations in flue gases themselves. Under these circumstances a community MWI would naturally shut down in community interest. A regional incinerator committed to a 24 h/d production schedule might be more reluctant to shut down unless a monitoring system blows a whistle.

To reach a decision to permit or not to permit based upon quantitative emission, dispersion, cost, and health risk analysis, we must develop the science of medical waste incineration much further. However, it should be clear that the motivation to apply pollution prevention for a community or local hospital MWI would be much greater than for a regional MWI.

Table 9–11 gives some useful approximate formulas for MWI discussions and estimations. Table 9–12 gives heating values for wastes, fuels, and plastics.

9.9 THE DECLARATION

The Declaration of the International Union of Air Pollution Prevention Associations (IUAPPA), approved on September 4, 1991, at Seoul, Korea,

TABLE 9–11. Some Useful Approximate Formulas and Conversion Constants

h = heating values (in MBtu/lb)   C = 12, carbon   H = 1, hydrogen
O = 16, oxygen   S = 32, sulfur   X = HCl = 36, hydrogen chloride

$[i] = m_i/m_t$ $C_aH_bX_cO_dS_e$ $h = 14.1[C] + 61.0[H] + 4[S] - 7.6[O] + 1[X]$

e.g., PVC $= C_2H_2X$, $[C] = (2 \times 12/62), [H] = (2/62), [X] = (36/62)$

waste-fuel mixture $h_b = (m_w/m_b) h_w + (m_f/m_b) h_f$, $m_b = m_w + m_f$

Combustion products: $[CO_2] = 3.66 [C]$, $[H_2O] = 8.94[H] + [H_2O^a]$
    $[N] = 8.86[C] + 26.4[H] + 3.3[S] - 3.3[O] + [N^a]$
$[HCl] = [X]$, $[SO_2] = 2[S]$

stoichiometric air-fuel ratio $= m(sta)/m_b = 11.5[C] + 34.3[H] + 4.3[S] - 4.3[O]$
    $\approx 0.75 \ h_b$ for most HCs

$T(°F) = \{3,000h_b/[1 + 0.75h_b(1 + ea)]\}$,   with 10% loss operating temperature
where   ea = fractional excess air

Exhaust flow conversion to dscfm at 68°F

$acfmx[528/(T_s + 460)] \times [1 - f(H_2O)] = dscfm$

Normalization to 7% O_2 Multiply by $(21 - \%O_2)/14$

Normalization to 12% CO_2 Multiply by $\%CO_2/12$

Residence time in secondary $t_r(sec) = 60 \ vol/acfm$

Conc.$[\mu g/Nm^3] = 0.2675$ EFxFR/dscfm, when EF in ppb, FR in lb/h

Max. burning rate PCC $B_r = 5 \ h_w^{1/2} \log Cap$ (lb/h)/ft^2

Heat release rate PCC $15 - 20$ (MBtu/h)/ft^3

Conversion Constants
M = 1,000   1 MBtu/lb = 0.556 kcal/g = 2.326 kJ/g
1 Btu = 1,055 kJ = 252 cal   1 lb = 0.4536 kg = 453.6 g
1 ft = 0.3048 m   1 ft^2 = 0.0929 m^2 1 ft^3 = 0.0283 m^3

[a] Denotes fraction in fuel.

and signed by J. Langstrom, Director General, on behalf of their executive committee is much stronger and broader than the pollution prevention program now being implemented in the United States. This resolution reflects the efforts of 28 member organizations representing 30 countries.

The Air and Waste Management Association (AWMA) represents the IUAPPA in the United States. It might be noted that AWMA's Medical Waste Committee is formulating a position paper whose first policy is to "promote the substitution of materials to reduce human exposure to toxics" (Donald Drum, private communication). The Declaration provides a vehicle for reconciling the divergent viewpoints on waste incineration as summarized in Section 9.8. The concepts defined in the Declaration also explicitly provide a broad framework for the topics of this book. The

TABLE 9–12. Heating Value of Wastes, Fuels, and Plastics in MBtu/lb

Component	As Rcvd.	Dry	Component	Dry
WASTES			**FUEL**	
Paper and Paper Products			*Hydrocarbons*	
Paper, mixed	6.80	7.57	Hydrogen	60.99
Newsprint	7.97	8.48	Natural	22.00
Brown paper	7.26	7.71	Methane	23.90
Trade magazines	5.25	5.48	Propane	21.52
Corrugated boxes	7.04	7.43	Ethane	22.28
Plastic-coated paper	7.34	7.70	Butane	21.44
Waxed milk cartons	11.33	11.73	Ethylene	21.65
Paper food cartons	7.26	7.73	Acetylene	21.50
Junk mail	6.09	6.38	Naphthalene	17.30
			Benzene	18.21
Domestic Wastes			Toluene	18.44
Upholstery	6.96	7.48	Xylene	18.65
Tires	13.80	13.91	Naptha	15.00
Leather	7.96	8.85	Turpentine	17.00
Leather shoe	7.24	7.83		
Shoe, heel, and sole	10.90	11.03	*Oils*	
Rubber	11.20	11.33	No. 1 (Kerosene)	19.94
Mixed plastics	14.10	14.37	No. 2 (Distillate)	19.57
Plastic film	—	13.85	No. 4 (VL Residual)	18.90
Linoleum	8.15	8.31	No. 5 (L Residual)	18.65
Rags	6.90	7.65	No. 6 (Residual)	18.27
Textiles	—	8.04		
Oils, paints	13.40	13.40	*Alcohols*	
Vacuum-cleaner dirt	6.39	6.76	Methanol	10.26
Household dirt	3.67	3.79	Ethanol	13.15
Food and Food Waste			**PLASTICS**	
Vegetable food waste	1.79	8.27		
Citrus rinds and seeds	1.71	8.02	Polyethylene	19.73
Meat scraps (cooked)	7.62	12.44	Polystyrene	16.45
Fried fats	16.47	16.47	Polyurethane	11.22
Mixed garbage I	2.37	8.48	Polyvinyl chloride	9.78
			PVC (pure resin)	7.20
Coals			Polyvinylidene chloride	4.32
Low-vol. bituminous	—	15.55	Polycarbonate	13.31
Med.-vol. bituminous	—	15.35	Cellulose	7.52
High-vol. bituminous	12.25	14.40	Polypropylene	20.02
Subbituminous	9.90	12.60	Polyester	12.81
Lignite	7.30	11.45		
Anthracite	—	14.00		

Declaration has been sent to all national governments and will be considered at the United Nations Conference on Environment and Development (UNCED), to be held in Brazil in June 1992. In view of its importance and the likelihood that it will soon receive international sanction (possibly with minor modifications), the Declaration has been reproduced, with the approval of the Director General, in full in the Preface of this work.

REFERENCES

Almaula, S. "Design and operation of thermal treatment units with waste stream modification." Paper 91-33.6 in the Proceedings of the 84th Annual Meeting of Air and Waste Management Association, Vancouver, British Columbia, June 16–21, 1991.

Aronson, D., and F. Hasselriis. Public perceptions versus the realities of solid waste management: The myths that grow and persist. Draft, August 1, 1991.

Barton, R. G., G. H. Hassel, W. S. Lanier, and W. R. Seeker. State-of-the-art association of medical waste thermal treatment. Report for the U.S. EPA Contract #68-0303365 and California Air Resources Board Contract #A832-155, Energy and Environmental Research Corporation, Irvine, CA, 1990.

Basic Environmental Engineering, Inc. The modern basic incinerator success story. 1990. Glen Ellyn, IL.

Brewer, P. "Air toxics monitoring program and emissions assessment for pathological incinerators." Paper 91-35.4 in the Proceedings of the 84th Annual Meeting of the Air and Waste Management Association, Vancouver, British Columbia, 1991.

Bulley, M. M. "Medical waste incineration in Australasia." Paper 90-27.5 in the Proceedings of the 83rd Annual Meeting of the Air and Waste Management Association, Pittsburgh, June 24–29, 1990.

Bulley, M. M. "The impact of design, procurement, and disposal options on medical waste incineration in Australasia." Paper 91-33.3 in the Proceedings of the 84th Annual Meeting of the Air and Waste Management Association, Vancouver, British Columbia, June 16–21, 1991a.

Bulley, M. M. "Incineration of medical wastes: Treating the cause rather than the symptom." Clean Air. May 1, 1991b.

Cassitto, L. "Destruction of organo-chlorinated micropollutants in combustion processes." 7th Miami International Conference on Alternate Energy Sources, 1985.

Chang, T. "Implementation of regional biomedical waste incineration facilities." Paper 91-30.6 in the Proceedings of the 84th Annual Meeting of the Air and Waste Management Association, Vancouver, British Columbia, June 16–21, 1991.

Cohen, Y., and D. Allen. "An integrated approach to process waste minimization research." Paper 91-43.5 in the Proceedings of the 84th Annual Meeting of the Air and Waste Management Association, Vancouver, British Columbia, June 16–21, 1991.

Deland, M. 1991. An ounce of prevention . . . After 20 years of cure. *Environ. Sci. Technol.*, 25(4): 561.

Dhargalkar, P. "An integrated waste management plant in Italy." Paper 91-43.6 in the Proceedings of the 84th Annual Meeting of the Air and Waste Management Association, Vancouver, British Columbia, June 16–21, 1991.

Doucet, P. E. "State-of-the-art hospital and institutional waste incineration: selection, procurement, and operations." Technical Document Series 055940, American Society for Hospital Engineering, Chicago, IL, 1991.

Ehrenfield, J. R., L. E. Susskind, E. P. Craig, and J. Nash. "Waste incineration: confronting the sources of disagreement—analysis of advocacy positions." Paper No. HSMP-14, Working Paper Hazardous Substances Management Program, Center for Technology Policy and Industrial Development, Massachusetts Institute of Technology, Cambridge, November 1988.

Glasser, H., and D. P. Y. Chang. "Analysis of the state of California's biomedical waste incinerator database." Paper 90-27.3 in the Proceedings of the 83rd Annual Meeting of the Air and Waste Management Association, Pittsburgh, June 24–29, 1990.

Green, A. ed. "Greenhouse mitigation." FACT-Vol. 7. American Society of Mechanical Engineers, ASME-H00513. 1989a.

Green, A. "Ozone depletion and greenhouse warming." In Proceedings of Woods Hole Conference, October 23, 1989b.

Green, A. Foreword to *Advances in Solid Fuel Technologies,* A. E. S. Green and W. E. Lear, Jr., eds. Fuels and Combustion Technology (FACT) Division of the American Society of Mechanical Engineers, FACT-Vol. 9, 1990.

Green, A., C. Batich, J. Wagner, and J. Blake. "Advances in uses of modular waste to energy systems." In *Advances in solid fuels technologies,* ibid. Green 1990a.

Green, A., et al. "Toxic Products from co-combustion of institutional waste." Paper 90-38.4 in the Proceedings of the 83rd Annual Meeting of the Air and Waste Management Association, Pittsburgh, June 24–29, 1990b.

Green, A., J. Wagner, C. Saltiel, and M. Jackson. "Pollution prevention and institutional incineration." Presented at American Society of Mechanical Engineers Solid Waste Processing Conference, Detroit, MI, May 17–20, 1992.

Hall, M. J., D. Lucas, and C. P. Koshland. 1991. Measuring chlorinated hydrocarbons in combustion by use of Fourier transform infrared spectroscopy. *Environ. Sci. Technol.* 25(2): 260–67.

Hasselriis, F. "Relationship between waste composition and environmental impact." Paper 90-38.2 in the Proceedings of the 83rd Annual Meeting of Air and Waste Management Association, Pittsburgh, June 24–29, 1990.

Hasselriis, F., D. Corbus, and R. Kasinathan. "Environmental and health risk of medical waste incinerators employing state-of-the-art emission controls." Paper 91-30.3 in the Proceedings of the 84th Annual Meeting of the Air and Waste Management Association, Vancouver, British Columbia, June 16–21, 1991.

Hasselriis, F., D. Drum, K. Martin, D. Brock, E. Schoettker, and A. Zanella.

"The removal of metals by washing of incinerator ash." Paper 91-32.2 in the Proceedings of the 84th Annual Meeting of the Air and Waste Management Association, Vancouver, British Columbia, June 16–21, 1991.

Hasselriis, F. "Effect of waste composition and charging cycle on combustion efficiency of medical and other solid waste combustors. For presentation at the 15th National Waste Processing Conference, Detroit, Michigan, May 18–20, 1992.

Jordan, J., C. Konheim, and J. McGrane. "Segregation of specific regulated medical waste items can be the centerpiece of medical waste reduction and cleaner incineration." Paper 91-33.7 in the Proceedings of the 84th Annual Meeting of the Air and Waste Management Association, Vancouver, British Columbia, June 16–21, 1991.

Kuntz, F., and G. Gitman. "Improved hazardous waste incineration with oxygen." Paper 91-19.3 in the Proceedings of the 84th Annual Meeting of the Air and Waste Management Association, Vancouver, British Columbia, June 16–21, 1991.

Lauber, J. D., and D. A. Drum. "Best controlled technologies for regional biomedical waste incineration." Paper 90-27.2 in the Proceedings of the 83rd Annual Meeting of the Air and Waste Management Association, Pittsburgh, June 24–29, 1990.

Lee, C., and G. Huffman. "Review of federal/state medical waste management." Paper 91-30.9 in the Proceedings of the 84th Annual Meeting of the Air and Waste Management Association, Vancouver, British Columbia, June 16–21, 1991.

Lee, C. C., G. L. Huffman, and R. P. Nalesnik. "Medical waste management: the state-of-the-art." U.S. EPA, Office of Research and Development, Cincinnati, OH, 1991.

Lerner, B. "The Beco Alka/Sorb process application to biomedical waste incinerators." Paper 91-30.1 in the Proceedings of the 84th Annual Meeting of the Air and Waste Management Association, Vancouver, British Columbia, June 16–21, 1991.

Licis, I. J., H. Skovronek, and M. Drabkin. 1991. Industrial pollution prevention opportunities for the 1990s. EPA/600/S8-91/052, U.S. Environmental Protection Agency, Cincinnati, OH.

Maxwell, G. "Performance of dry lime scrubbers with medical waste incinerators using field experience to correct design deficiencies." Paper 91-30.11 in the Proceedings of the 84th Annual Meeting of the Air and Waste Management Association, Vancouver, British Columbia, June 16–21, 1991.

Mineo, R. W., and A. Rosenthal. "Start-up and testing of a hospital waste incinerator." Paper 90-27.1 in the Proceedings of the 83rd Annual Meeting of the Air and Waste Management Association, Pittsburgh, June 24–29, 1990.

Oppelt, T. E. 1987. Incineration of hazardous waste: a critical review. *J. Air Pollution Control Assoc.* 37: 558–86.

Pasek, R., and D. Chang. "Potential benefits of polyvinyl chloride and polyvinylidene chloride reductions on incinerator emissions." Paper 91-33.1 in the Proceedings of the 84th Annual Meeting of the Air and Waste Management Association, Vancouver, British Columbia, June 16–21, 1991.

Rice, J., and G. Hester, "Risk management methodology for effective pollution prevention." Paper 91-45.10 in the Proceedings of the 84th Annual Meeting of the Air and Waste Management Association, Vancouver, British Columbia, June 16–21, 1991.

Riley, J., R. Knoche, and J. Vicinus. "Removal of heavy metals and dioxin in flue gas cleaning after waste incineration." Paper 91-35.9 in the Proceedings of the 84th Annual Meeting of the Air and Waste Management Association, Vancouver, British Columbia, June 16–21, 1991.

Saltiel, C., ed. 1991. *Impact of refuse-derived fuel on boiler design*. Vol. 13. New York: American Society of Mechanical Engineers.

Schifftner, K. "Integrating incinerator, scrubber, and waste disposal: The key to successful medical waste control." Paper 91-30.2 in the Proceedings of the 84th Annual Meeting of the Air and Waste Management Association, Vancouver, British Columbia, June 16–21, 1991.

Siebert, P., and D. Alston-Guiden. "Air toxics emissions from municipal, hazardous, and medical waste incinerators and the effect of control equipment." Paper 91-103.15 in the Proceedings of the 84th Annual Meeting of the Air and Waste Management Association, Vancouver, British Columbia, June 16–21, 1991.

Smolensky, L., S. R. Wysk, and C. E. Schmidt. "Development of an advanced control device for fine particulates." Paper 91-103.13 in the Proceedings of the 84th Annual Meeting of the Air and Waste Management Association, Vancouver, British Columbia, June 16–21, 1991.

Teller, A., and J. Hsieh. "Control of hospital waste incineration emissions—case study." Paper 91-30.5 in the Proceedings of the 84th Annual Meeting of the Air and Waste Management Association, Vancouver, British Columbia, June 16–21, 1991.

Theodore, L. 1990. *Air pollution control and waste incineration for hospitals and other medical facilities*. New York: Van Nostrand Reinhold.

U.S. EPA. 1990. EPA/530-SW-90-042. *Characterization of municipal solid waste in the United States: 1990 update*. Washington, DC: U.S. Environmental Protection Agency.

Wagner, J. "Control of modular incinerator to optimize destruction of toxics." Ph.D. diss., Department of Mechanical Engineering, University of Florida, Gainesville, 1992.

Walker, B. L., and C. D. Cooper. "Air pollution emission factors for medical waste incinerators." Presented at the 1991 Florida Section of the Air and Waste Management Association, Clearwater Beach, FL, Sept 15–17, 1991.

Weber, G. F., and G. L. Schelkoph. "Performance/durability evaluation of 3M Company's high-temperature Nextel filter bags." Presented at the 8th Symposium on Transfer and Utilization of Particulate Control Technology, San Diego, CA, March 20–23, 1990.

Wollschlager, M., and T. Casey. "Medical waste regulatory developments and facility compliance." Paper 91-28.2 in the Proceedings of the 84th Annual Meeting of the Air and Waste Management Association, Vancouver, British Columbia, June 16–21, 1991.

Index

209

DATE DUE